MELTDOWN

STORIES OF NUCLEAR DISASTER AND THE HUMAN COST OF GOING CRITICAL

JOEL LEVY

To Finn, Isaac and Michelle, for past meltdowns and future critical excursions.

Thanks to Issy Wilkinson at Welbeck, for cooking up the idea with me and guiding this project, and to Alison Moss for sympathetic editing.

The publishers would like to thank the following sources for their kind permission to reproduce the pictures in this book

Plate photographs in order of appearance: Science History Images/Alamy Stock Photo, Universal History Archive/Shutterstock, Atomic/Alamy Stock Photo, Los Alamos National Laboratory, Castle Light Images/Alamy Stock Photo, Katherine Jacobsen/AP/Shutterstock, Hulton-Deutsch Collection/CORBIS/Corbis via Getty Images, AP/Shutterstock, Everett Collection/Shutterstock, Dirck Halstead/The LIFE Images Collection via Getty Images/Getty Images, George D. Lepp/Getty Images, AP/Shutterstock, Media Production/Getty Images, Photo 12/Universal Images Group via Getty Images, Juliane Thiere/Alamy Stock Photo, Str/AP/Shutterstock, Graham Harries/Shutterstock, DigitalGlobe via Getty Images, Keystone/Zuma/Shutterstock

Every effort has been made to acknowledge correctly and contact the source and/or copyright holder of each picture and Welbeck Publishing apologises for any unintentional errors or omissions, which will be corrected in future editions of this book.

Published by Welbeck
An imprint of Welbeck Publishing Group.
20 Mortimer Street,
London W1T 3JW

First published by Welbeck in 2020

A CIP catalogue record for this book is available from the British Library

ISBN
Paperback - 9781787394995

Typeset by IDSUK (DataConnection) Ltd.
Printed at CPI UK

10 9 8 7 6 5 4 3 2 1

www.welbeckpublishing.com

CONTENTS

INTRODUCTION

On 14 December 1940, at a laboratory in Berkeley, California, scientists crowded around a machine that appeared to have lumbered free from the pages of a science fiction magazine. It consisted of two great cylinders, resembling torpedo tubes tapering to lance-like points, which fired infinitesimally small projectiles into a spiral channel trapped between two massive discs. These discs generated intense electromagnetic fields, which accelerated the particles to unimaginable velocities and smashed them against other particles with terrible violence. Led by future Nobel prize-winning American physicist Glenn Seaborg, the team of scientists activated the machine, throwing the switches that would unleash immense energies to forge within the strange crucible an entirely novel substance. The machine was a cyclotron – a particle accelerator or "atom smasher" – and from it Seaborg and his team would extract a minute quantity of something extraordinary, a new element from beyond the uttermost extremes of the Periodic Table.

It would be another 19 months until enough of the new "element 94" could be synthesized to gather a visible amount; months in which America had been plunged into a world war and begun to focus the full might of its scientific and industrial power on a top secret research programme to create the ultimate weapon. This programme, known as the Manhattan Project (see Chapter 1), would eventually lead to the design and construction of atomic bombs, but before this could happen it was necessary to invent new technologies and even new elements. To this end, Seaborg had now relocated to the Metallurgical Laboratory at the University of Chicago, and it was here, on 20 August 1942, that he recorded in his journal:

> today was the most exciting and thrilling day I have experienced. Our microchemists isolated pure element 94 for the first time . . . It is the first time that element 94 has been beheld by the eye of man.

By this time element 94 had a name, derived from the ancient Roman god of the Underworld: plutonium. In appearance plutonium is unremarkable, a silvery-grey metal that quickly tarnishes when exposed to air, but hold some in your hand and you will notice something strange; it is warm to the touch, for it continually generates heat. If you dropped a pellet of it on the surface of the miles-thick ice sheet at the South Pole, and left it for long enough, it would melt its way down to the bedrock. The source of this apparently inexhaustible energy is radioactivity, the phenomenon whereby unstable atoms spit out high-energy particles and electromagnetic rays. For a crash course in the science and terminology of nuclear fission, and a potted history of the development of nuclear science and the race to build the atom bomb, see the Appendix on page 355.

The dream of harnessing nuclear energy has led to the creation of a vast industry, which leverages the power of colossal machines and extreme engineering to tame the unruly energies of radioactivity and bend them to the generation of power for humanity. Yet plutonium was created to cause death and destruction on an unprecedented scale, and even before it was deployed in the heart of a bomb, it had been clandestinely injected into unwitting human guinea pigs, as part of a controversial programme of human experimentation with lethal consequences. Both edges of this nuclear sword were all too apparent from the very beginning.

This book is about the double-edged nature of nuclear power; its lurking threat, its unpredictability, and its potential for sudden and destructive outbursts, which can erupt more quickly than human thought. It is about the constant struggle to make safe the most dangerous force on Earth, the relentless effort to guard against every slip or unguessable calamity; and about the many, tragic times that this effort has failed. Some of these instances are grim milestones in global history, names and places that will reverberate for generations to come: Three Mile Island, Fukushima, Chernobyl. Some are barely known at all: Cecil Kelley, Hisashi Ouchi, Louis Slotin. In each case, this book recounts the human narratives of those caught in the disasters, and explains in clear and accessible language the scientific and technological details that led them there.

The story of nuclear accidents is almost as old as nuclear science. As soon as humankind had the ability to gather together in one place enough radioactive material for a nuclear chain reaction to occur, the danger loomed that such a reaction would run riot and smite those presumptuous enough to seek to control it. To the scientists engaged in these early experiments, such perilous visits to the edge of the nuclear precipice were known as "tickling the dragon's tail"; an acknowledgement that one misstep could lead to getting fatally burned. In essence, all subsequent attempts to manage nuclear power have been versions of tickling the dragon's tail, and in this book you will learn about all the ways in which that monster can be roused, and the consequences – human, environmental, financial and industrial – of unleashing its fury.

Chapter 1

EXPLORERS OF NEW WORLDS: 1944–45

IN THE late spring of 1944, a visitor to the train station of the small New Mexico town of Lamy might witness a procession of unusual arrivals. Dozens of highly educated young men and women, excited, tired after a long journey that might have brought them across an entire continent, and probably more than a little disoriented, would tumble off trains of the Atchison, Topeka and Santa Fe Railway to stand blinking in the desert sun. These visitors were directed to 109 East Palace Avenue in Santa Fe, some 16 kilometres (10 miles) up the road, and would make their way there to find a seventeenth-century, Spanish hacienda-style building located just off the main plaza. Inside, amidst a hubbub of noise and activity, they would be greeted by the friendly and reassuring presence of Dorothy Scarritt McKibbin. She would hand over passes and ID badges, assign living quarters and take charge of deliveries, and then send the new arrivals on their way.

By bus and automobile, the new recruits would wind their way along a 56 kilometre (35 mile) road that led up to an isolated mesa, the Pajarito Plateau, or, as its inhabitants knew it, "the Hill". When they arrived, they would see a small, white-painted clapboard shack bearing a large sign: "Los Alamos Project, Main Gate". Dorothy McKibbin, the First Lady of Los Alamos, dispatched an average of 67 people a day to the secret site on the Hill. Robert Oppenheimer, the man in charge, called her his "Gatekeeper", and relied on her to ensure the orderly flow of people and mail to the site codenamed Project Y, known today as Los Alamos National Laboratory after the small college and ranch compound that originally occupied this remote region of the New Mexico desert. Project Y had been established with one goal in mind: to produce an atomic bomb. The Los Alamos site had opened for business in early 1943, its population reaching around 3,500 by the end of the year, growing to 5,675 by the end of 1944, 8,200 a year later and 10,000 by the end of 1946.

During the Second World War, the top-secret research town had no official name or address; it did not appear on any map, and its only postal designation was a single PO Box, numbered 1663. All of the mail for the entire project came to this one address, and it was the only geographical information included on the birth certificates of the many babies born on the Hill during the Manhattan Project. So many baby baskets were ordered to this one PO Box that the delivery men of the main mail-order supplier, Sears, Roebuck & Co. were said to have become suspicious. Many of the boffins who worked at Los Alamos were science-fiction buffs, with the consequence that multiple subscriptions for the leading sci-fi pulp magazine, *Astounding Science Fiction*, were registered to this single address. In March 1944, the magazine's editor, John W. Campbell, came under investigation because a story he had overseen, "Deadline", was deemed to be suspiciously accurate in regard to technical details concerning uranium isotope separation. Campbell loved to tell the tale of how intelligence officers had grilled him in his office, oblivious to the fact that, pinned to the wall behind him, was a map of subscribers with a cluster of pins around an obscure location in New Mexico.

The new arrivals with whom McKibbin dealt included engineers, military police, soldiers, constructors and scientists of every field, from metallurgy and chemistry to mathematics and nuclear physics, ranging in age from 19 to 59. They included many of the greatest minds in science. Oppenheimer's weekly discussion groups might include up to six Nobel Prize winners, along with luminaries such as Richard Feynman, Edward Teller and John von Neumann. Talented men and women were brought together for a single overarching purpose, to build an atom bomb, but in order to achieve this they would have to develop new technologies and explore an entirely new world of science, one fraught with danger and unpredictable hazard.

Los Alamos was just one cog in the great machine of the Manhattan Project, which extended to vast industrial enterprises dedicated to the enrichment of uranium ore and the production of plutonium. The titanic labours pursued at enormous plants at Oak Ridge in Tennessee and Hanford in Washington state culminated in tiny quantities of precious radioactive core material, including U-235 and plutonium, being delivered to Los Alamos. Here there were teams working on two bomb designs; the gun-type design used to create Little Boy, the U-235 fission bomb that would be dropped on Hiroshima, and the implosion-type design of Fat Man, which would be dropped on Nagasaki.

Plutonium and heavily enriched uranium were effectively entirely new phenomena in the history of the Earth, let alone science, and many questions about the behaviour and response of such materials remained to be answered. So alongside teams perfecting the designs of the actual bomb devices, Los Alamos was also the site of feverish experimentation on the fissile material, as scientists sought to get a better understanding of what happens when fissile material goes critical, and the effects of different shielding, reflecting and tamper materials (tamper is a dense material, the inertia of which slows the expansion of reacting material to keep it together for longer, increasing the efficiency and yield of any explosion).

The first radiobiological accident on the Hill occurred in the summer of 1944. A young American chemist, Donald F. Mastick, was studying the chemistry of plutonium, which at that time was available only in minute quantities. On 1 August, Mastick was studying a vial of plutonium under a microscope. The vial exploded, spraying some of the highly toxic material into his mouth. It was estimated that he had swallowed about 10 micrograms of plutonium. Plutonium emits alpha radiation, which consists of alpha particles: relatively large, heavy, charged particles. These cannot travel far before smashing into something, making them easy to shield against (even a blade of grass will block an alpha particle). Their size, however, makes them an especially damaging form of ionizing radiation, since when a heavy alpha particle collides with an atom or molecule it is highly likely to damage it and knock off electrons. Alpha particles are 20 times more dangerous to human tissue than, for instance, gamma rays at the same dosage level. Thus, if an alpha-emitting plutonium particle can get close enough to biological molecules, it can do enormous damage to them, and this was precisely the danger facing Mastick. Even the very tiny amount he had ingested had the potential to wreak havoc on his delicate internal tissues, killing many cells and causing carcinogenic changes to others.

The director of the Health Group at Los Alamos was Louis Hempelmann, a young doctor from St Louis. The Health Group was the department in charge of what was euphemistically known as "health physics" – a brand new field covering radiation poisoning, exposure and dose regulations, and health and safety. Hempelmann would be a recurring actor in the radiobiological dramas that would unfold at Los Alamos. Called to the scene, Hempelmann promptly had Mastick's stomach pumped and his mouth heavily washed; he suffered no lasting damage and lived to the age of 87. The task of recovering the valuable plutonium from the stomach contents fell to the unfortunate Mastick himself. The incident was something of a wake-up call to Hempelmann's group; so little was known about the physiology and science of radiation exposure that Hempelmann himself called the state of radiobiology "primitive" and instigated an energetic programme of health physics research at Los Alamos.

One of the eminent scientists – many of them emigrés or even refugees from Europe – brought to Los Alamos was Otto Frisch, the Austrian-born physicist who had coined the term "fission" to describe the process that was the focus of the entire Manhattan behemoth. It was Frisch who, in conversation with his aunt Lise Meitner, had first explained the process of fission of the uranium nucleus in 1938 and calculated the extraordinary energies that it could unleash (see page 362). At the time, Frisch had been working in Copenhagen in the laboratory of Niels Bohr, becoming a leading expert in the physics of neutrons. With the outbreak of war, Frisch had found himself at Birmingham University in England, where he worked with Rudolf Peierls to produce the Frisch-Peierls Memorandum, the first document to set out exactly how a fission bomb might work and what the effects would be. Frisch had then worked for the British Tube Alloys project, the forerunner of the Manhattan Project, into which it was subsumed. Thus it was that in 1943 Frisch arrived in America as part of the British delegation sent to join Manhattan, and in 1944 he was posted to Los Alamos.

His engaging memoir, *What Little I Remember* (1980), gives something of the flavour of the time and the people at that extraordinary place. Frisch recalled, for instance, the division of labour at dinner parties, where he and other musically inclined scientists would wander in and start playing, evolving from solo performers to duet to trio as more people arrived. Meanwhile other scientists would be at work in the kitchen producing meals; occasionally the musicians would pause to consume a dish as it was served, before they returned to their instruments and the chefs to the kitchen. "That was repeated three or four times," Frisch recalled, "and the party ended around eleven o'clock, with well-filled stomachs and ears full of Beethoven and Mozart." With such a vibrant community of like-minded souls gathered at a remote location, there was plenty of opportunity for socializing. Frisch was inundated with offers, but his absent-mindedness caused problems: "I was once reduced to the expedient of sending a message over the paging system: 'Will the person who invited Otto Frisch to dinner please phone the number of his room'."

At Los Alamos, Frisch had the vital task of leading the Critical Assemblies group, and it was their job to discover exactly how much enriched uranium would be needed to make the core for the Little Boy bomb. The Frisch-Peierls Memorandum had supposed a critical mass of pure U-235, but obtaining pure U-235 was prohibitively difficult and inefficient. Instead the plan was to take uranium ore, which, in its natural state, typically contains little more than 0.7 per cent U-235, and enrich it to concentrations of 80 per cent or higher. With the help of reflector material (substances that would reflect escaping neutrons back into the reactive mass, thus increasing the probability of neutron capture by a fissile nucleus) it was possible to achieve criticality with a lower mass than would otherwise be needed, but the reflector layer added yet another variable to the complex calculus of criticality. The Critical Assemblies team worked to understand exactly how much of this enriched uranium would be needed to achieve critical mass with the help of reflectors, and how it would behave once it reached criticality. This was dangerous work; research on the very edge of disaster, and Frisch himself was one of the very first to discover just how easy it would be to stray across that edge.

Another of the ingenious schemes devised to decrease the mass needed for criticality was to mix the enriched uranium metal produced by the Oak Ridge plant with hydrogen-rich material to make uranium hydride. The hydrogen atoms in such a compound acted as moderators, slowing fast neutrons so that they could be more easily captured for fission, and thus increasing the efficiency of the chain reaction. Frisch's team used small blocks of uranium hydride roughly 3.8 centimetres (1½ inches) long to build Jenga-like cubic arrangements, constructing by hand a simulated atomic bomb core. As the construct neared the critical mass needed to sustain a chain reaction, so the danger of working with it increased. But dividing the material into small blocks gave the experimenters fine control over the critical threshold; adding a block might push the assembly to the very brink of going supercritical, and it would be relatively easy to subtract the fraction of the assembly necessary to dip back below the threshold. This at least was the thinking of the scientists, but it was a classic instance of hubris. The American physicist Raemer Schreiber, who worked at Los Alamos and was in the room with Louis Slotin when he received a fatal dose of radiation from the "Demon Core" (see page 37–38), recalled of such assemblages:

> You get fooled by the fact that they [are on the brink of going] critical; everything's kind of interesting. The counters go up; you move something off and the counters go down. It's a tantalizing sort of business. But if you go past critical – I don't know what the multiplication rate is, but it's far faster than human reactions.

In late 1944, a student was helping Frisch with what he called "an unusual assembly", known as the Lady Godiva assembly because it was "naked", in the sense of comprising uranium hydride blocks without any reflector around them. The experiment involved stacking several dozen of the little bricks and measuring the neutron flux as it approached criticality: "this was a good way to test the reliability of our calculations". Safety measures were limited to "little red signal lamps", which blinked every time the neutron counting meter registered a neutron. As Frisch placed more blocks onto the little pile, the neutron flux increased, and he and the student "both watched the little red signal lamps blinking faster and faster and the meter clattering with increasing speed". To his surprise, the meter suddenly stopped. Looking up, Frisch saw that the student had unplugged it. Leaning forward, he called out to the young man, "Do put the meter back, I am just about to go critical." This was truer than he realized. By getting closer to the assembly, Frisch had himself become a reflector; the hydrogen atoms in his body reflected back enough of the neutrons escaping from the assembly to tip it over the edge of criticality. "At that moment," he recalled, "out of the corner of my eye, I saw that the little red lamps had stopped flickering. They appeared to be glowing continuously. The flicker had speeded up so much that it could no longer be perceived."

Thinking fast, Frisch swept his hand across the top of the assembly to knock off some of the hydride blocks, "and the lamps slowed down again to a visible flicker". He realized immediately what had happened, and although he had felt nothing, running the radioactivity counter over some of the blocks showed that their activity "was many times larger than what should have accumulated if that little incident hadn't occurred". Frisch calculated that, for the two seconds in which his body had been acting as a reflector, the reaction rate had been increasing by around a hundred times a second. "Actually the dose of radiation I had received was quite harmless, but if I had hesitated for another two seconds before removing the material (or if I hadn't noticed that the signal lamps were no longer flickering!) the dose would have been fatal." Frisch's assessment that the dose was "quite harmless", when even by the lax standards of the wartime era he had absorbed in two seconds a full day's permissible dose, was typical of the somewhat cavalier attitude to radiation in the early years.

This attitude would again be demonstrated in the next, more serious incident to go awry at Los Alamos. On 4 June 1945, researchers were running an experiment to recreate the conditions that might obtain if a mass of enriched uranium were to be submerged in water, a genuine risk given that Little Boy was to be deployed in the Pacific Ocean region. Hydrogen-rich water acts as a moderator for neutrons, which is why many nuclear reactors feature cores sitting in water. Moderators increase the efficiency of fission and thus lower the critical threshold, so that dunking a sub-critical mass of enriched uranium might well risk sending it critical, effectively turning it into a nuclear reactor.

To test how a borderline-critical assembly would react to being submerged in water, 35.4 kilograms (78 pounds) of 83-per-cent-enriched uranium cubes were stacked inside a plastic shell and placed inside a steel tank. On top of them was a polonium-beryllium source – an external neutron generator to act like a starter motor for the fission process. Pipes running in to the bottom of the tank allowed water to be fed in and drained. As the experiment began, according to a 1971 review prepared by the Los Alamos Scientific Laboratory, "the immediate supervisor was absent from the scene". The two men operating the water valves began to fill the tank. As the water reached the level of the top of the polonium-beryllium source, the neutron counters showed that the rate of neutron production began to rise. Initially this seemed in line with expectations, but within seconds the count began to increase "at an alarming rate". As the *Manhattan District History* (the official record prepared by the Manhattan Project itself) laconically recorded, "The critical condition was reached sooner than expected". At that precise moment the experiment supervisor returned, walked within a metre of the tank and noted a tell-tale blue glow surrounding the tank. Fortunately for him and the other two men, they had already switched the valves so that the water began to drain out of the tank. The building was immediately evacuated, and the three men were taken away for observation. Two of them were assessed to have received radiation doses of about 66.5 rem, while the other had received about 7.4 rem. For comparison, the International Commission on Radiological Protection recommends an annual limit of 2 rem per year averaged over any five-year period, while 10 rem is the standard five-year limit for nuclear industry workers in the West today. See Appendix for more on rem and other units of radiation exposure.

According to the Los Alamos review, "no untoward symptoms appeared", while the *Manhattan District History* remarked merely that, "No ill effects were felt by the men involved, although one lost a little of the hair on his head." By the lackadaisical standards of the day, radiation-linked hair loss was apparently not considered "untoward". The uranium core itself became too radioactive to be used for a few days, but other than this the research team got off lightly.

Indeed, at this stage it was possibly the case that working with radioactive material was one of the less lethal pastimes at Los Alamos. Among the roughly two dozen tragic but comparatively mundane fatalities to be expected on a wartime project involving large-scale construction, there were some more unusual deaths, including two that were recorded as "accidental shootings", two from "self poisoning" (presumably suicide) and three from drinking ethylene glycol, possibly as a result of drinking "moonshine" (homemade spirits). Alcohol was officially prohibited on the Hill, which surprised Otto Frisch when he found out because of the ubiquity of boozing among the well-paid scientists. He personally resolved to quit drinking for good after catching himself absent-mindedly pouring a glass of tequila one morning. "Much later I learned it was against the rules to bring liquor to Los Alamos," he recalled, "but everybody did."

Frisch's run-in with a critical assembly might have daunted other souls, but instead he upped the ante, contriving a way to run experiments on assemblies that genuinely and deliberately crossed the threshold into criticality. By early 1945, he recalled, "enough uranium-235 . . . had arrived on the site . . . to make an explosive device". Frisch wanted to test his team's calculations and predictions with a scenario that was "as near as we could possibly go towards starting an atomic explosion without actually being blown up". Accordingly, he came up with an ingenious experimental design. A near-critical mass of uranium hydride would be assembled, but with "a big hole so that the central portion was missing; that would allow enough neutrons to escape so that no chain reaction could develop." The idea was to drop the "missing part", also known as the slug, through the middle of the doughnut-shaped assembly. As it fell, "for a split second there was the condition for an atomic explosion, although only barely so". Today this set-up is known as a pulsed fission reactor, since it produces only a pulse of criticality.

The frame that was constructed to hold and drop the plug was known as the "guillotine", coining one moniker for the experiment, but it is best known as the Dragon experiment, thanks to the physicist Richard Feynman. Despite his youth, the future Nobel laureate Feynman was on the council of senior scientists who assessed submissions for experiments on the Hill. Frisch recalled that Feynman received the proposal "with a chuckle" and described the audacious experiment as being "like tickling the tail of a sleeping dragon". Speaking many years later to a gathering held to honour his work on the Dragon experiment, Frisch likened his motivation to "the curiosity of the explorer who has climbed a volcano and wants to take one step nearer to look down into the crater but not fall in!"

Dangerous experiments at Los Alamos were conducted at remote sites; Frisch's groups worked at a laboratory in Omega Canyon. Here they built a 3 metre (10 foot) iron frame – the guillotine – that held in place vertical aluminium guide rails. At tabletop height, they built a ring of blocks of uranium hydride around the guides. Raised to the top of the guillotine was a slug of the same material, 5 by 15 centimetres (2 by 6 inches) in size. A complex series of fail-safes held the slug in place until its fall was triggered. The button that released the slug was labelled "HWG", for "Here We Go". The faster the slug was travelling as it passed through the assembly, the shorter would be the burst of fission reactions and neutron generation. Shorter bursts would be more favourable for research, and Frisch had considered accelerating the slug to 70,000 centimetres/second (2,300 feet/second) by firing it from a gun. "However," his original planning notes for the experiment record, "the use of artillery would introduce considerable complications and is not at present contemplated."

There are other disarmingly frank touches to Frisch's notes, which give something of the buccaneering flavour of early Los Alamos experimentation. For instance, discussing the dangers of the experiment, Frisch notes that, "One advantage of hydride is that an explosion, if it should happen, is less disastrous than with metal. We believe, however, that the arrangement can be made so safe that an explosion is humanly impossible." Later he muses on "possible causes why such a system might blow up". Perhaps most alarmingly, Frisch is remarkably casual about the worst-case scenario:

> Finally, it should be remembered that the reactor never gets more than about 0.1% super critical and that even if the slug did get stuck at the center the explosion would only be equivalent to a few tons of TNT. It would no doubt destroy the labs, but if this were built a mile from other laboratories these would be safe.

The chances of something going awry must have been greatly heightened by the pressure under which the scientists were working, as they raced to complete their experiment in the short window of time available before they had to hand the precious U-235 over to the armourers to be used in the preparation of the first atomic bomb. "During those hectic weeks," Frisch wrote in his memoirs, "I worked about seventeen hours a day and slept from dawn till mid-morning."

In the event disaster was avoided. Frisch himself claimed that, "Everything happened exactly as it should." The official Los Alamos history records that:

> These experiments gave direct evidence of an explosive chain reaction. They gave an energy production of up to twenty million watts, with a temperature rise in the hydride up to 2°C per millisecond. The strongest burst obtained produced 10^{15} neutrons. The dragon is of historical importance. It was the first controlled nuclear reaction which was supercritical with prompt neutrons alone.

According to some sources, however, the guillotine Dragon experiment pushed dangerously close to the edge of safety. It is recorded in the 1971 Los Alamos review as the first example of a critical excursion, because towards the end of the experiment, as the team sought to generate bursts of increasing power, a blast of 6 quadrillion neutrons "blistered and swelled the small cubes comprising the assembly matrix".

In one fascinating addendum to his notes, Frisch discusses the challenges of processing the large amounts of data that his experiment will generate and suggests constructing "an electronic adding machine" to process and synthesize data in binary form – in other words, an early form of electronic computer. In fact, John von Neumann, one of the fathers of computer science, was at Los Alamos with Frisch, who in his memoirs vividly recalled discussing with him the theory of computers and the possibility of building one with vacuum tube technology. Such encounters are illustrative of the thrilling intellectual foment of

Los Alamos, where the world's brightest minds had been brought together to tackle cutting-edge problems with every resource available.

Even the most brilliant minds could not entirely mitigate the risk of working on the edge of criticality. Frisch recalled: "Critical conditions could be reached very suddenly as the result of a minor mistake", while the in-house record of the atom bomb project, the *Manhattan District History*, noted that the criticality experiments were "especially dangerous" because "there is no absolute way of anticipating the dangers of any particular experiment, and the experiments seem so safe when properly carried out that they lead to a feeling of overconfidence on the part of the experimenter."

It seems likely that it was such overconfidence that led to the shocking death of Harry Daghlian, the first man to be killed by a critical excursion. Daghlian was an Armenian-American from Connecticut with a passion for particle physics, which had brought him to the laboratory of Marshall Holloway at Purdue University. When Holloway was asked to come to Los Alamos, Daghlian followed him. He was just 23 years old when he arrived on the Hill in 1944, and was assigned to Frisch's Critical Assemblies group, making an immediate impression with his size and strength. Daghlian's work included helping with the experiments on criticality, and in particular he worked with the plutonium cores that were being prepared for implosion devices. He was involved in readying the plutonium core that was used in the Trinity test, and he can even be seen in photographs and film footage of the team assembling the Gadget – the explosive device used in the test. In one clip we see a jovial looking, thickset man ambling across the front of the camera, having helped to load the plutonium core of the Gadget into, incongruously, the back seat of a beaten-up old Army sedan.

While the wartime aims of the Los Alamos effort culminated in the successful Trinity test, the bombing of Hiroshima and Nagasaki and the end of the War in the Pacific, the work there was just getting started. For scientists like Harry Daghlian, work continued at a breakneck pace, and some combination of pressure, overconfidence and misplaced bravado could lead to mistakes. On 21 August 1945, Daghlian was at the remote Omega Site laboratory, conducting "dragon's tail" criticality experiments with a 6.2 kilogram (13 pound) plutonium core, nicknamed "Rufus", intended to be used in the third implosion device (the first two being the Trinity device and Fat Man), which was to be detonated in the first of a series of postwar atomic tests. The experiments involved testing the neutron flux produced by surrounding the core – a sphere of polished metal sitting in a machine-cut cradle block – with differing configurations of reflector that could improve the neutron economy (see above). In this case, the reflector material was bricks of tungsten carbide. By hand, the experimenter would stack the tungsten carbide bricks around Rufus and measure the neutron flux that resulted as the assembly neared criticality.

As Otto Frisch later recalled, there were two basic safety rules for those working on such experiments: "nobody was to work all by himself, and nobody should ever hold a piece of material in such a way that if dropped it might cause the assembly to become critical". Having observed both rules for several hours of working on the core with colleagues, Daghlian returned from a dinner break and promptly broke both: "He was so eager that he wanted to do one more assembly". After stacking five layers of bricks, so that they almost surrounded the core, Daghlian was moving to place a brick in the centre of the assembly when his monitors warned him that the reaction rate was cresting alarmingly, and that this last piece would send the assembly critical. As he pulled back his hand, the heavy tungsten brick slipped from his fingers, falling directly on to the top of the core. "Even as he instantly swept it aside with a blow of his muscular arm," related Frisch, "he saw a brief blue aura of ionized air around the assembly." Daghlian had dropped the brick from his left hand and used his right to knock it off. He reported feeling a tingling sensation in that hand, at the same time as he saw the blue flash.

Frisch would later have pause to reflect once more on that blue flash, when recounting how he had witnessed the first atomic explosion, the Trinity test of 16 July 1945, in the desert near Alamogordo, a place also known, Frisch noted, as *El Jornada del Muerte* (The Journey of Death). Just before dawn, parked up about 40 kilometres (25 miles) from the test site, Frisch was woken from a doze by the countdown to the test. Unable to find his goggles he sat with his back to the explosion and witnessed a sudden light as though "without a sound, the sun was shining . . . The sand hills at the edge of the desert were shimmering in a very bright light, almost colourless and shapeless." Once the glare had dimmed sufficiently, he was able to watch the fireball and its cloud of smoke and dust rise from the ground, which initially seemed to Frisch to "look a bit like a strawberry" and then 10 seconds later like "a red hot elephant standing balanced on its trunk". But, "as the cloud of hot gas cooled and became less red, one could see a blue glow surrounding it, a glow of ionised air; a huge replica of what Harry Daghlian had seen when his assembly went critical and signalled his death sentence".

What is the nature of this blue glow? It is generally assumed to be an ionized-air glow: the result of high energy radiation smashing into the electrons orbiting atoms in the air. Some of the collisions are strong enough to knock electrons off atoms altogether, producing ions (charged particles), and so this kind of radiation is often described as ionizing radiation. When electrons absorb bursts of energy like this, they are said to become excited, leaping from lower to higher energy states, and when an excited electron drops back down to its "resting" state (de-excites), it emits a packet of energy in the form of a photon. Some of these photons are in the wavelength of visible light, and the colour emitted depends on the nature of the elements being de-excited. Atmospheric air is primarily composed of nitrogen (78 per cent) and oxygen (21 per cent), so in air, the predominant contributor to the resulting spectrum is ionized nitrogen, which emits lots of blue wavelength photons when de-exciting. This is thought to be the main reason that critical excursions and atomic explosions alike produce an eerie blue glow or flash. De-exciting nitrogen and oxygen ions also emit lots of infrared photons, which are not visible but may be sensed as radiant heat, and it is thought that this might account for the "heat wave" reported by some critical excursion victims.

Ionized-air glow is different from Cherenkov radiation, which is a related phenomenon whereby deceleration of high-energy particles in a fluid medium produces blue light. Ionizing radiation produces a Cherenkov glow only in relatively dense fluids such as water, which is why nuclear fuel elements stored in water pools often give off a characteristic blue glow. Accordingly many texts warn against confusing the blue flash of criticality with Cherenkov radiation. But there is an unsettling possibility that this phenomenon may well be responsible for the blue flash seen by excursion victims such as Daghlian, and his many unfortunate successors, from Louis Slotin (see page 26) to Alexander Zakharov to Hisashi Ouchi (see page 305). One theory is that the blue flash seen by people in such close proximity to a burst of radiation actually comes from inside their own eyes; from the Cherenkov radiation produced as ionizing radiation passes through the dense medium of the humours in the eyeball. (Another consequence of ionizing radiation affecting air is that the oxygen ions produced react together to form ozone [O_3], an unusual allotrope with a distinctive smell. Workers involved in the clean-up of Chernobyl reported that the smell of ozone was a sign of high radiation levels.)

Apart from the tingle in his hand, Daghlian reported no immediate symptoms, but even as the ambulance took him to hospital he began to vomit. According to the Los Alamos *Review of Criticality Accidents*, during the brief second that the plutonium core went supercritical there were around 10^{16} fission events, and Daghlian received a dose of 510 rems (just over the 500 rem threshold normally deemed fatal). The plutonium cores were priceless government property, and accordingly each was accompanied at all times by an armed guard. So in fact Daghlian was not completely alone; the core's guard, who was in the same building but not close to the experiment, received a dose of 50 rem. At the time there was little extant expertise around assessing the dose of radiation received in such an event; poor Daghlian was about to become a textbook case. Metal articles that he had about his person, such as coins and a belt buckle, were collected and sent to be assessed for their levels of induced radioactivity. Radioactivity levels of such items correlated to the amount of energy they had absorbed and thus gave clues to the size of the overall dose Daghlian had received. Radiation biologist Dr Wright Langham came by to pick up the items.

Over the next two weeks Daghlian succumbed to the effects of acute radiation sickness, aka acute radiation syndrome (ARS). The hand that he had used to knock away the tungsten bricks was the worst affected; it developed blisters and sores. Fast-dividing cells in the body are the most vulnerable to radiation damage (this is the premise of radiotherapy to treat cancer), and bone marrow cells fall into this category. Bone marrow is responsible for producing many of the white blood cells that make up a vital part of the body's immune system, and a catastrophic fall in the white blood cell count – the number of white blood cells circulating in the blood – is a tell-tale symptom of ARS. This in turn fatally compromises the sufferer's immune response, making them unable to fight off even normally harmless pathogens. As Frisch summarized, "two weeks later, his blood count way down, [Daghlian] died from some trivial infection which his body could no longer fight." In fact, Daghlian suffered for 24 days.

Daghlian was just 24 years old. Still only a PhD candidate, he was not married, and so the authorities sent for his mother. Marguerite Schreiber, whose husband, the nuclear physicist Raemer Schreiber, was away at the time in the Pacific as part of the team that had assembled the Fat Man bomb, was asked to look after her. Raemer Schreiber recalled that his wife was told nothing about the accident and still didn't know anything years later, which suggests perhaps that Daghlian's mother was not told much either. In a 1993 interview with historian of the atom bomb, Richard Rhodes, Schreiber himself revealed a rather unsympathetic take on the tragedy: "It was just stupid . . . because there were rules which were clearly violated. Somebody gets killed, you had to go around and say he was stupid."

Others were more sympathetic. Louis Slotin was a colleague of Daghlian; both had been part of the team that assembled the Trinity test Gadget. Slotin was called upon to help estimate the dose of radiation Daghlian had received, working with Wright Langham to assess the induced radioactivity of the metal items taken from Daghlian. Slotin was even tasked with writing the official Los Alamos investigation, "Report on Accident of August 21, 1945 at Omega Site". But he was also a friend of the stricken man, sitting with him for many hours during the long weeks it took him to die. Did he feel a shiver of presentiment at any point?

Los Alamos as an institution did try to learn some lessons. There was a review of safety regulations and the establishment of a special committee to review criticality experiments and oversee safety procedures. The procedures were formalized and strengthened to include the following safeguards: requiring a minimum of two researchers; use of at least two monitoring instruments for neutron flux, each equipped with audible alerts; more rigorous planning of operating procedure; and contingency planning for problems or emergencies. Yet Harry Daghlian was just the first name on the tragic roll call of those who lost their lives to critical excursions, his accident a chilling foreshadowing of a much worse mishap that would involve the very same plutonium core.

Chapter 2

LOUIS SLOTIN AND THE DEMON CORE, 1946

THE ATOMIC bombing of Hiroshima and Nagasaki marked the culmination of the Manhattan Project and the achievement of the goal towards which the assembled intellectual might of Los Alamos had been working. News of the successful detonation at Hiroshima provoked contrasting emotions. Otto Frisch recalled vividly, "the feeling of unease, indeed nausea, when I saw how many of my friends were rushing to the telephone to book tables at the La Fonda hotel in Santa Fe, in order to celebrate." He understood that "they were exalted by the success of their work" but could not forget "the sudden death of a hundred thousand people". The moral qualms of many would be exponentially exacerbated by the second bombing, just a few days later. Had it been necessary? Wouldn't the Japanese have surrendered anyway?

In the days and months to come, many of the scientists involved in the bomb project would wrestle with a profound burden of guilt. Perhaps the superstitiously inclined believed that their work would exact a karmic toll. Perhaps there was even a sense that some form of expiation was needed. The agent of this karmic retribution had a taste for destruction, for it had claimed one life already; it would come to be known as the Demon Core – the same plutonium sphere that had delivered a fatal dose of radiation to Harry Daghlian. Its next victim would be a scientist who was representative of so many of the young scientists who arrived on the Hill, whose relatively short career had encompassed in microcosm the entire Manhattan Project, and whose unique experience made him better suited than almost any other to understand exactly what had happened to him, and what would happen next.

Louis Slotin was a Canadian physicist and biochemist, born in 1910 in Winnipeg, of Russian-Jewish stock. Although his father worked in livestock, his studious son was clearly not destined to follow in his footsteps. Young Louis proved to be a brilliant chemist with a gift for experimentation. At the precocious age of 16 he went to study biochemistry at the University of Manitoba, developing into a young man with "a romantic and elaborate view of himself and the world", as a friend later recalled. It was at this time that he adopted the middle initial "A", for reasons of style apparent only to himself. After gaining his doctorate in biochemistry in London in 1936, he returned to North America, where his studies began to change direction.

Slotin became fascinated by the pioneering cyclotron being developed at the University of Chicago. This was an early atom-smashing device, which accelerated charged particles to colossal speeds before colliding them to produce new atomic nuclei or reveal subatomic particles. Slotin joined other enthusiasts to help construct the machine, scavenging copper wire and blowing glass tubes himself. For three years, from 1937 to 1940, he worked for free since there were no salaried positions available, prompting his exasperated but proud father to ask, "What have I, a student prince on my hands?" When a friend asked him what kind of doctor Louis was, Slotin senior responded by flicking a light switch on and off. "Do you know where the light went to? You don't know. I don't know. But my son Louis knows. That's the kind of doctor he is." Louis was known for being reserved but self-possessed, with an antic sense of humour; he would amuse himself, for instance, by seeding among gullible colleagues false beliefs about a romantic past, such as that he had fought in the Spanish Civil War or flown with the RAF.

Slotin worked on a project that used the cyclotron to create an unusual isotope of carbon, which in turn could be used as a tracer in experiments on photosynthesis. In 1942 the Manhattan Project came to the University of Chicago, with the setting up of the Metallurgical Laboratory under Arthur Compton, the American physicist, and Enrico Fermi, the Italian Nobel laureate who had fled to America to escape Fascist persecution of his Jewish wife. Fermi had been struggling to achieve the first sustained fission chain reaction in a pile – an assemblage of blocks of uranium and graphite. But he was also heavily involved in work on plutonium (having won his Nobel prize partly for research into the element), and it was this that was the main focus of the Metallurgical Laboratory. Slotin, meanwhile, through his work on radiobiology, attracted the notice of Manhattan Project recruiters, and was tapped to work under future Nobel laureate Eugene Wigner, on the production of plutonium.

Slotin was likely to have been present at the successful activation of the world's first nuclear reactor, the forerunner of all subsequent reactors and the first achievement of a controlled chain reaction. Fermi and his team had been experimenting with different combinations of uranium metal, uranium oxide and graphite (an allotrope of carbon, vital for moderating the neutrons liberated by the fission of uranium nuclei – see Appendix). At Chicago Fermi and others had drawn up plans to build a large pile on the university campus, but Compton and other higher-ups at the Manhattan Project had understandable misgivings about doing such an experiment in the middle of a metropolis. Instead it was decided to set up a research institute in the Argonne Woods Forest Preserve, about 32 kilometres (20 miles) from Chicago, but when construction at the site was delayed by an industrial dispute, Fermi convinced Compton that he could be trusted not to blow up the city. Accordingly Chicago-Pile 1 (CP-1) was assembled in a squash court under the stands next to the university's football field.

On 2 December 1942, the Metallurgy team and other invitees gathered on the balcony of the squash court to watch history being made. On the floor of the court sat the pile of uranium and graphite blocks; control rods, made from sheets of cadmium wrapped around wooden sticks, kept it from going critical. Arthur Compton described the scene:

> Fermi gave the order to withdraw the control rod another foot. We knew that that was going to be the real test. The geiger counters registering the neutrons from the reactor began to click faster and faster till their sound became a rattle. The reaction grew until there might be danger from the radiation up on the platform where we were standing. "Throw in the safety rods," came Fermi's order. The rattle of the counters fell to a slow series of clicks. For the first time, atomic power had been released. It had been controlled and stopped. Somebody handed Fermi a bottle of Italian wine and a little cheer went up.

Among the 49 scientists present was, according to some accounts, Louis Slotin. The reactor had run for about 4½ minutes at a power of just 0.5 watts; at such low power it had not been deemed necessary to install any shielding, so that, as Compton had recognized, those viewing from the balcony were exposed to the neutron flux being generated. It was perhaps Slotin's first brush with a potentially dangerous dose of radiation.

Slotin would soon (probably around April 1943) move on to a different part of the Manhattan Project, travelling as part of Eugene Wigner's team to Oak Ridge, Tennessee, where the world's second nuclear reactor, the X-10, was being used to transmute uranium into plutonium. At Oak Ridge a daring but disturbingly reckless side of Slotin emerged. According to a tale recounted by one of his Oak Ridge colleagues, health physicist Dr K. Z. Morgan, Slotin was impatient to make adjustments to an experiment at the bottom of the tank of water that helped shield against radiation from the reactor. It was a Friday afternoon and he wanted the reactor shut down so that he could make the changes but was told he would have to wait. When Morgan got back to work on Monday morning, he recalled, "I found that Louis had stripped down to his shorts, dived into the tank and made the adjustments under water. I was appalled that anyone would take such risks. It shows what kind of person he was. He was like a cowboy – but a good experimental scientist."

From Oak Ridge Slotin was recruited to Los Alamos, to join Robert Bacher's Weapon Physics Division, aka the G (Gadget) Division. This was the department tasked with developing and building the core of the implosion-type plutonium device, which would later be used for both the Trinity Gadget and Fat Man, the bomb that would be dropped on Nagasaki. Slotin arrived on the Hill in December 1944, and was not impressed with what he found, describing it as a "disorganised mess".

G Division had been formed when it became clear that the plutonium produced at Oak Ridge was not suitable for use with a gun-type device. Work on the alternative technology, the implosion-type device, had been stalled, but now it assumed vital importance. Slotin became an integral part of the team racing to realize this demanding technical challenge, one of the most difficult aspects of which was working with plutonium, a quixotic and highly toxic material. Only in April 1945 were sufficient quantities of plutonium finally available for G Division to begin in earnest their preparation of the bomb "pits": the term used to describe the inner portions of the bombs, resembling the hard core of an apricot or peach, and comprising the fissionable core and the hard shell of tamper and neutron reflector around it. Tamper is dense material with high inertia that helps to slow the explosive expansion of the core, so that fissioning nuclei remain close to each other for longer, thus increasing the yield of the bomb. (Often the material used for the tamper – depleted uranium – was also a neutron reflector, and so accounts of cores and pits at the time often use the term "tamper" in place of "reflector".)

Over the next three months the team worked hard to determine the most efficient configuration of plutonium and tamper for a critical assembly in the pit, while the many other elements in the complex implosion design came together around them. With the White House pushing hard for a test as early as possible, the whole Manhattan Project raced to be ready. Originally the deadline had been set as 4 July, but this was already deemed unrealistic by March.

Los Alamos considered that a revised deadline around 23 July was achievable, but White House pressure led to a commitment to aim for a weather-dependent window of 15–16 July.

Slotin acquired a reputation as a skilful assembler of bombs, especially adept at combining the elements of the bomb's firing mechanism. As the day of the Trinity test approached, he assumed a central role in the drama. In photos and footage of the lead-up to the test, Slotin can be seen in the midst of the action, his slight figure ostentatiously cool in an unbuttoned shirt and shades. Last-minute experiments, and delays to explosive components of the Gadget device, made 15 July the earliest achievable test date, but this was to be preceded by several days of practice assembly of the bomb. Accordingly on 11 July the core – packed in a specially designed "field case" – was loaded into the back of an Army sedan by scientists including Slotin and Harry Daghlian, and driven 340 kilometres (211 miles) south to the test site at the Alamogordo bombing range.

At the McDonald ranch house, a small wooden building 3 kilometres (2 miles) from the tower where the device would be detonated, Slotin and other members of the team gathered to practise assembling the bomb pit. The high-explosive elements of the Gadget arrived on site on the 13th, and that same day the final assembly of the device began. This marked the point at which the core was formally passed over to the Army to be blown up, and it was to Slotin that the receipt for the precious plutonium was made out. It was signed by Brigadier General T. F. Farrell, deputy for General Leslie R. Groves, the head of the entire Manhattan Project. Afterwards, Slotin would keep this unique document in the glove compartment of his cream 1942 Dodge Custom Convertible Coupe.

For the final stages of the assembly, the team moved all the equipment to a canvas shelter erected at the base of the test tower. Photos show Slotin next to the Gadget, looking relaxed, even casual, but as the official *Manhattan District History* records, this was the first time that "active material in large quantity was put within high explosives . . . Although the people performing the operation and those watching it were outwardly calm, there was a great feeling of tension apparent." By 17.00 the Gadget was assembled and hoisted to the top of the tower, 33 metres (110 feet) up. People dispersed to various observation posts, including the main control bunker situated roughly 9 kilometres (10,000 yards) away.

The *Manhattan District History* does not record where Slotin positioned himself for the actual test. Many of the scientists parked up a few miles from the tower and prepared for the initial blast by lying down with their feet towards the explosion. Slotin presumably wore his sunglasses, and was spared the fate of Hans Bethe, the German-American physicist, who looked directly at the blast and was completely blinded – fortunately, only for a short while.

The Trinity test was quickly followed by the successful detonation of the Little Boy and Fat Man bombs over

Hiroshima and Nagasaki, and the subsequent surrender of Japan. Victory in the Pacific marked the end of the Second World War, and almost as soon as the hostilities were over, many of the Manhattan Project scientists left Los Alamos to return to academia. Slotin, however, was not able to extricate himself quite so easily; he was too valuable. "I am one of the few people left here who are experienced bomb putter-togetherers," he lamented.

Los Alamos in the immediate aftermath of the war was a much-diminished place. The rush for the exit, combined with problems with the quality of living accommodation at the site, which came into sharper focus without the overriding impetus of the war effort, led to a severe decline in morale. Slotin's main focus, however, was on the urgent push for further tests of America's new superweapon. The devastation of Hiroshima and Nagasaki bore ample testimony to its destructive potential, but many questions remained unanswered. Naval chiefs, in particular, wanted to know what effects such a weapon might have on ships at sea, and they petitioned the President for permission to conduct a series of marine tests. The series was given the go-ahead for the summer of 1946. "It was apparent," said Vice Admiral W. H. P. Blandy, head of the test series task force, "that warfare, perhaps civilization itself, had been brought to a turning point by this revolutionary weapon." Accordingly he named the test series Operation Crossroads.

The original plans for Crossroads called for use of three of the plutonium cores that the Manhattan Project had by now accumulated. Louis Slotin was one of the senior scientists left at Los Alamos who, having taken over from Otto Frisch as leader of the Critical Assemblies group, had the experience and expertise to work with these cores in the lead up to the tests. With Crossroads scheduled to start in July, Slotin spent the months leading up to the test doing criticality experiments on the core nicknamed Rufus – the same core that had killed his friend Harry Daghlian in August of the year before. Slotin was scheduled to fly out, with the Rufus core, to the Bikini Atoll in the South Pacific, and assemble the bomb for one of the Crossroads shots. This would mark the end of his Manhattan Project service; afterwards he would take up a teaching position at the University of Chicago. By mid-May he had already packed up 11 crates of books and other stuff and sent them on ahead.

Slotin's replacement on the Critical Assemblies team would be Alvin Graves, a physicist from Washington, D.C. Like Slotin, Graves had been part of the Metallurgical Laboratory at Chicago – he had helped construct CP-1 – and had later been transferred to Los Alamos. Now he was to be transferred to the Critical Assemblies team to take over from Slotin, and on 21 May he was one of the participants at a group leaders' meeting at Slotin's laboratory, the Omega Site in Pajarito Canyon, 6.5 kilometres (4 miles) from the main compound.

The day had begun for Slotin in familiar fashion, in his bachelor's rooms at the main site, where he dressed in his typical attire: an expensive sports shirt and khaki slacks, tucked into a pair of cowboy boots. He had had lunch with a friend at the Technical Area PX, before making the journey to the Pajarito Omega Site for the meeting. The meeting concluded with a tour around the Omega Site, including the building housing the laboratory where Slotin's team, and another team led by Raemer Schreiber, performed their critical assembly experiments. The laboratory was a large, bare room, containing little fixed furniture but cluttered with equipment, including racks of radiation counting equipment, pallets of lead blocks for different shielding configurations, and a small, low metal table in the centre, on which the near-critical masses were assembled. The visiting party inspected the lab and departed, with the exception of Graves. He and Slotin had been discussing an assembly that Graves had never seen in action. "Why don't I run through it for you now?" asked Slotin, setting his empty Coca-Cola bottle down on the table. According to a memo penned by Los Alamos director Norris Bradbury, a few days later, this demonstration may have been part of the final checks for the core at the heart of the assembly, which was being prepared for its transport to Bikini.

It was around 15.00, and those present in the room besides Slotin and Graves were three men from the laboratory staff – Allan Kline, Marion Cieslicki and Dwight Young – and Private First Class Patrick Cleary. Cleary was the security guard whose job it was to accompany at all times the valuable plutonium core nicknamed Rufus, soon to become better known as the "Demon Core". Also in the lab, engaged in an experiment of their own over by the east wall, were Schreiber and his assistant, 23-year-old Theodore Perlman. As Schreiber later recalled, "that afternoon we had to check some initiators, which were in the same laboratory".

Rufus, the subject of the experiment, was a small polished metal sphere, 89 millimetres (3½ inches) in diameter and weighing 6.2 kilograms (14 pounds), made of a plutonium-gallium alloy, plated with a thin layer of nickel. The tiny fraction of gallium in the alloy helped to stabilize, thermally and chemically, the highly reactive plutonium, while the nickel plating protected it from oxidization, blistering and corrosion. The amount of plutonium in the core was sub-critical (though warm to the touch, heated by its own radiation); it would be brought to the very limit of criticality by enclosing it in a shell of beryllium tamper. In fact, the beryllium was there as a neutron reflector, but was called a tamper in the argot of the day. The core itself was already set in the lower half of the tamper shell: a thick hemisphere of beryllium, itself set within a still larger hemisphere of polished aluminium, from which the top half of the core's sphere projected, like the pit of a plum when half of the flesh has been cut away.

To match the lower half-shell of beryllium, there was a 23 centimetre/9 inch-diameter upper half-shell with a hemispheric void in the centre, so that it could be placed over the projecting plutonium half-sphere and complete the tamper shell. To fully complete the assembly, however, would increase the efficiency of neutron capture in the plutonium to such a degree that it would go critical, and so a number of steps were taken to prevent this. The upper half-shell also had a hole through it, so that it could be held by putting a thumb through the hole, like a bowling ball. This allowed basic manual control of the positioning of the upper half-shell. It was also standard to position spacers – inert blocks such as slivers of wood – around the upper face of the lower shell, to ensure that the two shell halves could never meet. Indeed if the two half-shells came closer than within 3 millimetres (⅛-inch) of one another, enough neutrons would be available to trigger prompt criticality, generating a runaway fission chain reaction for the fraction of a second it would take for thermal expansion of the assembly (thanks to the intense heat generated) to cause it to expand and no longer be dense enough to be critical. Though lasting only a millisecond, this burst of prompt criticality would generate a blast of neutrons, gamma rays and beta particles.

It was the risk of such a burst that made the criticality experiment so dangerous; it was another iteration of the "tickling the dragon's tail" experiments that had started with Otto Frisch (see pages 14–15). Slotin was, by now, an old hand at dragon tickling, but it was evident to many that he was almost literally playing with fire. Notoriously, Slotin had been warned by Fermi, "If you keep on doing that you'll be dead in a year." Frisch recalled that Slotin "told me that Fermi warned him, 'You know that in this sort of work you have perhaps an even chance to survive your work here.' Slotin was rather shaken about it."

With Graves set to take over, this was the last time that Slotin would be tickling the dragon's tail. Perhaps he wanted to put on a good show for his replacement; perhaps familiarity had bred carelessness. Eyewitness accounts agreed that, having started the demonstration in regulation style, Slotin then began to push his luck. He began by taking hold of the upper half-shell of beryllium with his left hand, with his thumb through the hole and his fingers splayed out around the top of it. The spacers were in position, and Slotin moved the upper shell closer to the lower one; as the gap between them narrowed, the activity of the core picked up. The increasing neutron flux was picked up by the counters and registered as an increasingly urgent clicking noise and a climbing red trace on a roll of graph paper. So far, so routine.

But now Slotin removed the spacers, replacing them with the blade of a screwdriver, which he gripped with his right hand. Colleagues interviewed about this action years later gave slightly different accounts. One described this development as "something different – not extraordinary but not routine". Another suggested that the assembly was not running as normal, "So Slotin improvised". Still others insisted that Slotin followed normal procedures.

With his left hand, Slotin lowered the side of the upper shell furthest away from him to rest on the lower half, while his right hand held the screwdriver in place as the last line of safety. By shuffling the screwdriver blade tiny increments further back, he was able to vary minutely the gap between the two shells, sending the neutron counters into ever more intense frenzies of clattering. It was 15:20; Graves, standing just behind Slotin, leant round slightly to get a better view. Behind him and to the left stood Cieslicki of the lab staff. A metre or so away from the other side of the central table were the other two lab staffers, Kline and Young, and, further back, the security guard, Cleary. Perlman was with Schreiber, who, busy with his own work, had turned his back on the dragon demonstration.

Graves heard a click as the screwdriver blade slipped out of the gap and the two beryllium half-spheres closed together. Schreiber heard the same noise and turned round to see a flash of blue light and feel a wave of heat on his face. In a report he wrote a week later, he recalled:

> The blue flash was clearly visible in the room although it (the room) was well illuminated from the windows and possibly the overhead lights . . . The total duration of the flash could not have been more than a few tenths of a second. Slotin reacted very quickly in flipping the tamper piece off.

Schreiber was not the only one to comment on Slotin's speedy reactions. A Los Alamos *Review of Critical Accidents*, contrasting Slotin's accident with Daghlian's, notes that in this case "the experimenter was better prepared to disassemble the material, and it is thought that this was done in a fraction of a second, perhaps < ½ second." It was also widely put about that Slotin had instantly and bravely thrown himself over the core, to shield his fellow scientists. In fact, it hadn't mattered how quick his reactions were or how swiftly he moved to shield his colleagues; the reaction had been stopped by thermal expansion far quicker than any human reflexes could act. For the brief millisecond that it went on, however, the reaction produced about 3 quadrillion fission events; a million times fewer than the Trinity test or Nagasaki bomb, but quite enough to generate a lethal blast of radiation.

Everyone in the room would have known what had just happened; all except one man, PFC Cleary. As he later recalled:

> Our instructions [were] to keep in sight of all active material that is around, except in the case of a critical assembly, but [I] am not sure about that. I did not actually know what the material or sphere was at the time, or anything about it . . . When the accident occurred, I saw the blue glow and felt a heat wave. I knew something was wrong, but didn't know exactly what it was, when I saw the blue glow and somebody yelled . . .

Cleary, Young, Cieslicki and Kline turned on their heels and ran. They bolted out of the east door of the laboratory, and Cleary, Cieslicki and Kline ran to the Pajarito compound's gate and hustled the MPs on duty to open it. They and the MPs then ran up the road for a short distance. Young had stopped to take cover behind an earth barricade, but when Slotin did not come out of the critical assembly building he went back and looked in through the door. Seeing no one, he circled round to the main laboratory building, north of the assembly building, where he found Slotin and the others. They had followed Perlman, who had run out of the assembly room, along a corridor to the main labs.

There Slotin displayed remarkable sangfroid. Recounting the incident many years later, Schreiber was still impressed by Slotin's reaction: "He knew he had had it at that time. He kept his head." Schreiber himself, "went back around the corner with a meter and that went off scale, so I decided I better not go back in. I guess [Slotin] called the hospital and told them to send the ambulance out." Slotin also marshalled the group who had fled up the road to come back to the lab, and drew up a sketch that showed roughly where everyone had been standing at the moment of the critical excursion. Then he called his friend Philip Morrison, a physicist who had been at the Metallurgical Laboratory in Chicago, had worked with Slotin on the Critical Assemblies team, had helped assemble the Fat Man bomb in the Pacific, and had visited Hiroshima and Nagasaki in the aftermath of their destruction as part of the damage assessment team. Morrison had seen at first hand the effects of massive radiation exposure. He remembered the call from Slotin:

> Lou said: "We've had an accident. It went prompt critical and you'd better come down here. I've called the hospital." Then either he said or I asked, "There was a blue glow." We both knew that was very bad.

"Just how bad?" was the overriding question in everyone's mind. The radiation dose that each man had received would determine his fate, and the affected men scrambled to estimate their exposure. As they waited for the ambulance, they made a number of confused attempts to measure radiation levels. One man tried to scan the assembly lab with a radiometer that was already in there, only to realize that it was contaminated and simply registering its own radioactivity. None of the men had been wearing dose badges (which used photographic film to record radiation exposure), and Slotin asked Perlman, who had been furthest away from the core, to retrieve the badges from the locked lead box in which they had been stored, and throw them onto the assembly. It was a dangerous and pointless task, which was later cited in a report on the incident as evidence that, in the wake of such an incident, those affected "are in no condition for rational behavior."

Slotin's left hand told its own story. To begin with it was mostly numb with some tingling, but it would become increasingly painful. This was the part of his body that had been closest to the core when it went prompt critical, and it had been blasted with over 15,000 rem of X-rays, as well as other radiation. Slotin's combined whole-body dose for neutrons, X-rays and gamma rays was estimated at around 2,100 rem; 500 is normally considered a fatal dose. He was by far the worst affected; the other people in the room received doses estimated to range from 360 to 37 rem.

Another estimate put Slotin's exposure at 1,000 rads (recall that rads are a measure of the amount of energy absorbed, while rem is a measure of the biological damage sustained by absorbing that energy – see Appendix). This was an order of magnitude higher than Kline, who was just over a metre away from Slotin, while Graves, who had been standing just behind Slotin, received an estimated 166 rads. Whole body doses of more than 1,000 rad are almost always fatal, while doses between 100 and 200 cause acute radiation syndrome (ARS) but are not normally fatal.

Slotin, with a background in biochemistry, expertise in nuclear physics, and eyewitness experience of the consequences of acute radiation exposure, was perhaps uniquely placed to understand what had just happened to him, and what was going to happen next. He was hoping for the best, but he feared the worst. When he and Graves got a moment together, the first thing he said to him was, "Al, I'm sorry I got you into this. I'm afraid I have less than a fifty-fifty chance of living. I hope you have better than that."

Although they had called for an ambulance, the affected men took matters into their own hands and drove themselves to the hospital. All except one; PFC Cleary, summoned back from his panicked dash beyond the site gates, had to wait outside the lab in which lurked the lethal core, still thermally and radioactively hot from its excursion. Hot or not, the core was US Government property of astronomical value, and Cleary would have to sit tight until he was relieved. His exposure had been relatively limited and the incident would have little impact on his military career. He would later be sent to fight in Korea, where he was killed in combat on 3 September 1950, just a few weeks short of his 25th birthday.

At the hospital, the first order of business was to gather clues to the men's radiation exposure. One way to do this was to measure directly the radiation counts of their bodily tissues, and of samples of blood and other bodily fluids. The intense blast of neutrons had rendered Slotin's body tissues themselves radioactive; later, emissions from radioactive sodium in his own cells would be measured and used as an indirect method of calculating the neutron flux to which he had been exposed.

Another way to estimate exposure was to measure the level of induced radioactivity in metal that the victims had about their persons. Accordingly, nurses collected from them items ranging from coins and fountain pens to pen knifes and silver belt buckles. At 18:00, Wright Langham came into the ward. "I know why you're here," Slotin told him; just as he had done nine months earlier, after Daghlian's accident, Langham had come to take away the metal items to be measured for radioactivity. For Slotin, history was repeating as tragedy.

Langham's health physicists, aka radiation biologists, knew precious little about acute radiation syndrome. They were, however, probably already aware of the grim conclusion they would reach in the review they completed after Slotin's death and autopsy: "tissues can be protected to some extent against the damaging action of penetrating radiations, [but] there is no known method of saving doomed cells once injury has occurred".

Slotin's next visitor was Philip Morrison. After visiting the scene of the accident at the Pajarito site, he dropped in to see his stricken friend. They talked about the dosage and Morrison asked if he could get anything for Slotin, who requested something to read. That evening Morrison went to the Los Alamos machine shop to get one of the engineers there to help rig up a book rack that could hold open a book and turn the pages individually by means of clips attached to strings, in turn attached to a ratchet that could be advanced by an elbow-operated switch. He understood all too well what was about to happen to Slotin's hands.

A photo taken very soon after Slotin's admission to the hospital shows his hands laid on top of the bed sheets, in a gesture of supplication, or resignation. Apart from blisters on the ends of two of his fingers, they look unscathed, although possibly a little puffy. But by 18:30, Slotin's left hand, which had been holding the beryllium half-sphere, was swollen and red, while the thumb on that hand, which had been projecting through the hole in the half-sphere, and had thus been exposed to the naked, unshielded core, was going numb and its nail bed blackened. His right hand had less severe symptoms, but by Wednesday afternoon, 24 hours after the accident, it too was severely swollen, while the skin on the left hand looked like it might split or burst.

The same photo also shows a thick bandage around Slotin's lower abdomen, the part of his body that had been level with the core. By Wednesday it too was beginning to swell. Apart from this, however, Slotin actually felt better than he had the day before, having stopped vomiting. Daghlian had displayed a similar progression of symptoms, and indeed remission of initial symptoms of sickness is a common feature of ARS, with sufferers often displaying relatively good spirits as a result. This remission is likely due to cells and tissues that did not receive an overwhelming dose of radiation rallying from the immediate insult, using their reserves of cell machinery and resources to carry on for a while. The true cost of exposure emerges when it comes time for cells to make use of their DNA – for production of new proteins, and most especially when it comes to replicating. Those cells that replicate most often are the "canaries in the coalmine" for ARS – hence the dramatic impact on bone marrow and blood cell production. However, those tissues most directly in the firing line, such as Slotin's hand and midriff, which were closest to the core, immediately began to suffer the impact of the catastrophic damage they had received, since they had literally been cooked, like meat in a microwave.

Not all those involved in the incident displayed this pattern of symptoms. Allan Kline, the 26-year-old physicist who had been on the other side of the table from Slotin, and had received a potentially life-threatening dose of radiation, was suffering badly. After experiencing nausea and vomiting on the first day after the accident, Kline suffered fainting spells, complete loss of appetite and rapid weight loss over the following few days. Like all those affected, however, he tried to put a brave face on things, insisting to the nurses that his vomiting was due to a combination of nerves and bad hot dogs, rather than radiation.

Schreiber and Perlman had been far enough away to be only mildly affected. Schreiber recalled that, after the accident, the two of them jumped in a jeep, "and I stopped off to my apartment and told my wife that I was going to have to go and be checked into the hospital, but I was all right. Well, I was thirty feet away." He remembered being detained for just "a day or so", during which "they took blood samples and stuff. But anyhow, it didn't bother my health, so I went on with going overseas."

As Wednesday wore on, the condition of Slotin's left hand worsened steadily. Eventually it would take on a waxy blue look, with huge blisters, initially on the thumb but then, by Thursday, on the palm and between the fingers. The rest of his arm swelled up, as did his right hand and forearm. Doctors packed both hands in ice and plied Slotin with morphine for the steadily worsening pain.

That day there had been a conference for scientists trying to work out Slotin's dose – and thus his ever-diminishing odds of survival. Langham's initial estimates, based on the metal items that had been taken off Slotin, including his watch, small change, a ring and a gold pen, were that the Canadian had received a dose four times higher than Daghlian's fatal exposure. Later he would recall:

> Being a relatively simple person, I didn't see how he had much of a chance. But the physics boys were still calculating, calculating, calculating. I walked in and told them what I thought. Phil Morrison picked up my data and pitched it the length of the desk. He said, "Hell. It can't be."

Morrison was even less receptive to the military's knee-jerk public relations response. After Daghlian's accident the Army had released a vague statement that gave little away, referring to an unspecified incident involving "technical personnel" who were now in a "satisfactory condition". A similar statement was now being drafted about the Slotin incident, but Morrison kicked up a storm, threatening to go to the newspapers if the Army simply covered things up. Accordingly, a remarkably open press release was issued admitting that personnel had been exposed to radiation, and by 25 May the incident had made the national newspapers. Further, equally frank press releases would follow after Slotin's death; such transparency would not prove to be the norm in the months and years to come (see below).

With news of the accident now in the public domain, Slotin was given permission to contact his family. He dictated a telegraph to be sent to his father: "My trip to Pacific indefinitely postponed, will write details love Louis." Later that Thursday, in the evening, a nurse held the receiver while he spoke on the telephone to his parents. Slotin told them he'd had a little accident and would neither be travelling to Bikini nor able to visit them at home afterwards, so perhaps they would like to come to see him. The Army arranged flights for Mr and Mrs Slotin, who set out the next day.

While his parents were en route, Slotin's physicians were getting creative in managing his worsening pain. As ice packs and morphine ceased to ameliorate the pain from his hands, the doctors moved to encase completely in ice his right hand and his left arm and hand, which had similar effects to amputation. He was receiving blood transfusions, with friends queuing to donate blood, and he remained alert and engaged, asking every physicist who visited, "Well, what's the dose?" Morrison visited to read to him, and colleagues brought flowers from their gardens. Another visitor was a photographer, who set up lights to take pictures in gruesome colour of Slotin's affected body parts, for some of which he patiently posed naked. Like Daghlian before him, and many victims since (see in particular the Tokaimura incident), Slotin was discovering that his personal tragedy was all too easily subsumed by the cold imperatives of scientists, hungry for data derived from a horribly exclusive set. Doctors collected the bodily fluids of Slotin and the others, day and night.

Slotin's parents arrived at Albuquerque on Saturday morning and were met by Philip Morrison. "Louis is my oldest son," Mr Slotin explained to Morrison on the drive to Los Alamos, "and every father loves his son. But there is more than that. There is respect for Louis, for a learned man." When they reached their son's bedside, he was sitting up and talking lucidly; they were fortunate to have arrived during the small window of latency before the full impact of radiation sickness was felt. Slotin tried to reassure them that he had received "just a bit of a burn", but his mother was disturbed by the feeling of his hair, exclaiming, "it's stiff and dry, like wire", while after seeing his son, Mr Slotin, evidently in need of a stiff drink, quietly asked Morrison where he could get a bottle of whisky. The Slotins were not the only arrivals; a Dr Hermann Lisco from Chicago had been called to Los Alamos. He was a pathologist, who might perform an autopsy when one was required.

On 26 May, Slotin's medical report recorded that, "From this day on, the patient failed rapidly". It was five days since the accident. His pulse and temperature began to fluctuate and his white blood cell count plunged. Annamae Dickie, the nurse with responsibility for blood monitoring, burst into tears when she did her routine count of white cells in Slotin's sample. It was obvious now that there was no hope for him. Nonetheless, Slotin retained enough coherence to suggest a fix for a painful ulcer on his tongue, which he realized was actually a radiation burn caused by his gold filling. The metal filling had become so radioactive that it was burning adjacent tissues. At Slotin's suggestion physicians wrapped the tooth in a protective shield of gold foil; the dense metal makes an excellent barrier to radiation. Kline was suffering from the same phenomenon, only in his case there was a chance that he might survive the ARS, and his doctors were anxious that radiation from metal inlays in his teeth might dramatically enhance his risk of cancer of the jawbone. An initial mouthpiece crafted from gold foil proved too thin to block the radiation, and it was replaced with a solid gold one, which Kline wore for five days, until the induced radioactivity of the inlays subsided.

Slotin suffered from increasing abdominal pain and lost weight rapidly. The doctors likened the internal radiation burns in his body to "three-dimensional sunburn". This phrase was even used in an official press release about the incident. Morrison, in a memo to colleagues about Slotin's progress, reported: "The fifth and sixth days were evidently very hard ones."

Slotin's condition continued downhill. His organs began to fail and his tissues began to fall apart. His skin turned a purplish-red, his abdomen became stiff and distended and his gastrointestinal tract completely broke down and had to be drained through his nose. His platelet count followed his white blood cells and dropped precipitously. Platelets are part of the blood's clotting system. From his experiences with the victims of Hiroshima and Nagasaki, Morrison knew that "this was a sure sign of the onset of the haemorrhagic phase" – the grisly denouement of ARS that saw victims bleeding from every orifice. Slotin had seen it too, in Daghlian. "Both Louis and I knew enough about this to be unhappy about its coming. It is likely the next four or five days would have been very unpleasant," Morrison wrote to Slotin's friends, reflecting that it was perhaps fortunate that Slotin began slipping in and out of consciousness, by Wednesday becoming delirious. His lips turned blue and he was put into an oxygen tent but by night-time he was in a coma. At 11:00 on Thursday, 30 May, Louis Slotin died; he was 35 years old. The cause of death was recorded as "acute radiation syndrome".

Morrison and Captain Paul Hageman, head of the Los Alamos hospital, sat down with Louis' father and told him that there would need to be an autopsy. Mr Slotin pointed out that it was against his religious beliefs, and that the Jewish community back in Winnipeg would criticize him for allowing it, but he gave permission nonetheless, out of respect for Louis' commitment to science. The Army flew the specially sealed coffin to Winnipeg, accompanied by his parents, in a converted transport plane from which the normal interior had been stripped so that a sofa could be bolted down for them. Louis Slotin was interred at Winnipeg's Shaarey Zedek Cemetery; more than 2,000 people attended his funeral. Today his life and tragic end are commemorated by a small park dedicated to his memory, close to his family home, at the end of Luxton Avenue on the banks of the Red River in north Winnipeg.

Meanwhile the business of Los Alamos continued, and the Demon Core had somewhere it was supposed to be. While Slotin was in hospital, the director of Los Alamos, Norris Bradbury, sent round an update to scientists involved in preparing for the upcoming Crossroads tests. The core had been getting "its final check", but evidently the situation had now been radically altered:

> Obviously Slotin will not come to Bikini. [Raemer] Schreiber will come although the date of special shipment was postponed one week to allow us to pull ourselves together. Only two shipments will be made at this time as I see no courier for the third. The sphere in question is OK although still a little hot but not too hot to handle. We will save it for the last in any event if it is needed at all.

Bradbury's statement gives the lie to the long-received understanding about the fate of the Demon Core, which in most accounts is said to have met a fittingly fiery and destructive end as the core of either the Able or Baker shots of the Crossroads tests. But as Bradbury acknowledged, the core was "a little hot" to be considered for either of the first two shipments (in other words, for either the Able or Baker shots), and accordingly was to be held in reserve for the third, Charlie shot. The Crossroads Operation, however, was terminated after Able and Baker, and so the Demon

Core never achieved its long-intended apotheosis in a nuclear explosion. So what did happen to it? Nuclear historian Alex Wellerstein discovered, while researching a comprehensive history of the Demon Core for the *New Yorker* magazine in 2016, that the third core went out with a whimper rather than a bang, as it was melted down and recast into subsequent cores for the atomic weapons programme. Before this, however, it was almost certainly used in recreations of the accident, which were diligently staged by the team investigating what had happened as they attempted to glean every morsel of evidence from the mishap.

Indeed the authorities – initially the Manhattan Project, then in its last gasp, and subsequently, after the signing into law of the Atomic Energy Act by President Truman on 1 August 1946, the Atomic Energy Commission (AEC) – went to extraordinary lengths to recreate the accident. Health physicists made hollow life-size models, filled with synthetic blood to simulate the bodies of the affected men. Known as phantoms, these models were placed around the laboratory in positions corresponding to Slotin and the others, while the accident was rerun using the same core, albeit with everything being operated remotely. After each run, the radioactivity levels of the phantoms' blood were compared with the blood samples of the real victims, to judge the fidelity of the simulation.

The immediate institutional impact of the accident was to bring a halt to criticality research at Los Alamos. A memo written in the aftermath of the incident recommended using remote controls in order to make "more liberal use of the inverse-square law" – the fact that the intensity of radiation drops exponentially with distance. Nuclear historian Richard Rhodes, in conversation with Raemer Schreiber in 1993, noted that "after this second accident, as I recall, Bradbury made a hard rule that all criticality tests would be run by remote control. Which they were."

Slotin was the only one of the affected men to die from ARS, though probably not the only fatality caused by the incident. Alvin Graves (who had been standing behind Slotin) died of a heart attack in 1966, which, according to a 1978 follow-up study on those affected in the accident, was probably caused by complications associated with radiation exposure. Two of the other affected men died of blood disorders that may have been linked to radiation. Remarkably Allan Kline, who, after Graves and Slotin, probably received the highest dose, had beaten the odds and recovered enough from his dose to be released from hospital two weeks after the accident. He was promptly fired, since regulations specified how much radiation you could be exposed to, and Kline had been exposed to more than two decades' worth even by the lax standards of the day. According to the regulations, he would not be allowed to work with radioactive materials for at least 25 years. Slotin's slip of the hand had brought his career in nuclear physics to an abrupt end.

For Kline, however, this was just the start of a long and bitter battle with the authorities. Right from the start, Los

Alamos and then the AEC dissembled, misled and tried to shirk their responsibility for Kline – or Case Five, as he was officially designated. While Kline was in the Los Alamos hospital, Norris Bradbury wrote to his mother to tell her that her son was "not seriously affected" and that he had "only minimal symptoms". This was despite his hospital chart stating plainly: "The depression of the lymphocytes and leukopenia [low white cell count] which developed makes it obvious that this man's exposure was significant. Final Diagnosis: Radiation Sickness."

Discharged from hospital and no longer able to work in the nuclear industry, Kline returned to his hometown, Chicago, his hair falling out, his immune system seriously affected and his health hanging by a thread. Louis Hempelmann, director of the Health Group at Los Alamos, aware of Kline's compromised immune system and enhanced risk of cancer, advised him to stay out of the sun for at least two years and to cover up from head to toe whenever he had to venture outside. Accordingly, on the rare occasions on which he had the energy to go out, Kline braved the summer heat dressed in a sombrero, long trousers and women's elbow-length gloves.

The follow-up care that Kline had been offered consisted of periodic visits to, initially, the Metallurgical Laboratory in Chicago, and later to the Billings Hospital. The former took 22 blood samples between June and November 1946, while at the Billings, Kline was subjected to a battery of tests when he went in on 8 December. It became apparent to him that the invasive testing regime to which he was being subjected was not for his benefit, and he stormed out of the hospital two days later. His anger mounting, Kline now attempted to claim compensation from Los Alamos and the AEC but found himself engaged in a protracted lawsuit. The authorities had long been expecting the suit and had been preparing to cover themselves. On 10 December, the same day that Kline walked out of Billings, Hempelmann wrote to James J. Nickson, medical director of the Argonne Medical Laboratory and Kline's physician in Chicago, that the affair was causing everyone at Los Alamos "a most remarkable case of jitters . . . This case is being handled in a most unusual manner. We [. . .] have been instructed not to contact Kline directly nor to commit the project in any way."

Over the next three years the Manhattan Project and then the AEC stonewalled Kline's efforts to obtain his own medical records or obtain significant acknowledgement of the harms done to him. For instance, in 1949 Kline's lawyer wrote to a senator who had sponsored the legislation that created the AEC, setting out in detail the health problems his client had suffered as a result of the Slotin incident, and complaining that both the AEC and the University of California, which operated the Los Alamos laboratory, were withholding his medical records. The AEC's response consisted of flat denials and outright lies: "Neither the files of the Commission nor the University disclose any request by Mr. Kline for such information nor any indication of any refusal by the University or the Commission to furnish such information to him." Instead, the AEC claimed, "With respect to the extent of the injuries sustained by Mr. Kline, the statements contained in [the lawyer's] letter . . . appear inconsistent with available medical reports."

The AEC even resorted to attacks on Kline's character, but let slip the true motive for their blanket denials, arguing that lawmakers "should consider the fact that there may be many other individuals . . . who have been exposed to radioactive emissions . . . and [support for Kline] might lead to a deluge of requests". In other words, they feared opening the floodgates for expensive claims against them, and with good reason. Kline's case was just the tip of the iceberg; of 24 people with a claim to compensation for injury resulting from exposure to radiation, not one had received the proper medical treatment, let alone compensation. The AEC refused to acknowledge, in any of the two dozen cases, that any claimant had indeed been injured by radiation.

Kline battled for many years to receive proper acknowledgement and compensation, but eventually his personal circumstances changed. He overcame his injuries and long-lasting health problems to forge a lucrative new career in IT, one which brought him back into the orbit of the military-industrial complex. Perhaps unwilling to bite the hand that now fed him, Kline retreated from the controversy. He died in 2001.

Were Kline and the other affected men angry or bitter about the accident, or towards Slotin for having caused it? Certainly Slotin's death provoked complex emotions among his colleagues in general; ones they seemed reluctant to examine. Barbara Moon, a Canadian journalist researching the Slotin story in 1961, found that many of those working at Los Alamos did not want to discuss the incident with her. One told her: "I don't wish to talk about him at all", while Philip Morrison told her: "It was the most painful time of my life and I don't like to go back to it." Speaking in 1993 about the deaths of both Slotin and Daghlian, Raemer Schreiber was not sentimental (see page 24). Moon herself attributed the ambivalence of Slotin's colleagues to a lingering sense of guilt about having helped engender the Atomic Age, which she cast in biblical terms: "it may be that if they must remember Louis Slotin they must also feel again what they felt in those first days after mankind lost his innocence."

Chapter 3

THE TOWN THAT WASN'T THERE: THE KYSHTYM DISASTER, 1957

DEEP IN the heart of USSR, just beyond the Eastern Urals, is the most radioactively polluted place in the world. Today a 70 kilometre/43 mile-long strip of land, officially designated a nature reserve, is still fenced off and prohibited to all but a handful of people because of its lingering radiation. In the 1990s parts of the site still blazed with radiation roughly a thousand times more fierce than the highest naturally occurring rates. Fish in a nearby reservoir were reported to be "100 times more radioactive than normal", while in 1991 children in a village down the river from the site were being exposed to chronic radiation at an effective dose of up to 1 rem/year. In the UK, for comparison, nuclear industry trainees aged under 18 are allowed a maximum annual exposure of just over half this. Lurking beneath a concrete cover is the world's most radioactive body of water, a cesspit of dumped radioactive pollution so virulent that as late as 1998 – and possibly still today – it was fatal to stand on its shore and visitors had to approach in a lead-lined tank. This is where, in 1957, at a site so secret that it did not officially exist, the Soviet Union managed to dirty bomb itself.

What is known to history as the Kyshtym Disaster actually occurred at the Mayak Processing Plant in Chelyabinsk-40 (known today as Ozersk), a town that could not be found on any map because its existence was a highly guarded state secret. In the immediate aftermath of the Second World War, the Soviet Union's overriding priority was to catch up with the atomic bomb technology of the Allies. All other considerations were secondary, and no heed was to be paid to the cost, particularly the cost in human lives; manpower was the one resource that the Soviets possessed in inexhaustible supply.

The Soviet A-bomb programme was jump-started by the information provided by German-British spy Klaus Fuchs. Fuchs was a German communist who had taken refuge in Britain in the 1930s, where he became a physicist and was recruited to work on the Tube Alloys project (the British precursor to the Manhattan Project). When the British effort was subsumed into the Manhattan Project, Fuchs moved to the USA and eventually came to Los Alamos, working at the heart of the team developing the implosion-style plutonium bomb. All the while he was feeding top-secret information to the Soviets. Fuchs was not uncovered until 1949, by which time he had provided an extraordinary wealth of information to the Soviets, credited with accelerating the development of the first Soviet atom bomb by up to two years.

Another helpful source of information about methods and facilities came in the form of a handy publication produced by the USA at the end of the War: *Atomic Energy for Military Purposes*, aka the Smyth Report. This invaluable document, available to any member of the public, included a map of the Hanford Works (the colossal facility where uranium was turned into plutonium), a clear photograph and even a guide to synthesizing plutonium-239. The Smyth Report was a bestseller in the USA, and was popular with Soviet engineers and scientists, especially after a Soviet translation was published in early 1946.

Taking Hanford as their model, the Soviets started work on a similar plant at Chelyabinsk-40 (known simply for its postcode, a sub-district of the nearest city, Chelyabinsk, east of the Urals and close to USSR's southern border), later known as Chelyabinsk-65, and today known as Ozersk. The plant and town were run by the Mayak (meaning "lighthouse" or "beacon") Production Association, and so the enterprise was known as the Mayak Chemical Combine. Work on the site had already started in November 1945, and 12 prison camps were emptied of 70,000 inmates to provide forced labour, toiling to build a complex that sprawled across 90 square kilometres (35 square miles). The first production reactor was brought on line in June 1948, transmuting uranium into plutonium for use in an implosion-style device, but already the Soviet A-bomb project was racing against the clock because of desperation among the top echelon to deliver the first test in time for Stalin's birthday in December. In the event the test, at the Semipalatinsk test site in modern-day Kazakhstan, was not delivered until August of the following year, but it was nonetheless a resounding success, shocking the Allies, who had assumed the Soviets were still years away from such an achievement.

Lack of equipment and know-how, combined with the intense secrecy around the project, meant that many of the thousands of workers had little idea what they were doing, and still less of the dangers involved. The information supplied by Fuchs or gleaned from the Smyth Report left many gaps, and the plant's designers and operators had to make up the rest as best they could. Constant pressure for results, and the breathtaking disregard for human life that was a feature of the Stalinist system, helped ensure that the safety record of the plant was appalling. Much of the work of the plant involved processing contaminated liquids to recover valuable plutonium. This was inherently dangerous – plutonium is highly poisonous and lethally radioactive, and many of the chemicals used in the processing were also toxic – but the crude methods involved exacerbated every part of the process. Untrained workers sloshed solutions around in buckets, mopping up spills with rags. Containers of deadly concoctions went unaccounted for, and protective gear was non-existent. Hundreds of workers suffered plutonium poisoning, while 1,500 more developed chronic radiation syndrome (CRS), a condition resulting from long-term exposure to sub-lethal doses of radiation. There were 500 cases in 1952 alone.

With few or no safety protocols, it was easy for critical masses of radionuclides to come together in vessels and go supercritical (i.e. achieve prompt criticality). There was a string of such incidents, including, for instance, one on 15 March 1953, in which a worker unknowingly added a concentrated solution of plutonium to a vessel already containing plutonium, which resulted in the vessel holding nearly a kilogram (2 pounds) of plutonium. The unfortunate man was unaware that the deadly concentration of radionuclide was spewing radiation at his legs. When the solution started boiling, workers divided the contents between two other vessels, and the unsuspecting victim went about his business. Two days later he was hospitalized with symptoms of acute radiation syndrome (ARS). It was estimated that he had been blasted with 1,000 rads, severely burning his legs, which had to be amputated. Two other men received less severe doses but needed treatment for ARS. Another accident later that year caused five cases of ARS. These were just the first in a long sequence of accidents that told of dismal safety standards, poor training, mismanagement and shoddy equipment, among other problems.

The baleful influence of the plant spread far beyond its immediate boundaries, since it was perhaps the most polluting factory in history. Production of plutonium was the only thing that counted, and for the first few years of operation no attempt was made to deal with the colossal amounts of contaminated water and other waste produced. Instead it was dumped into the Techa river, a small, meandering waterway that eventually flows into the Tobol, itself a tributary of the mighty Ob, a river that discharges into the Arctic Ocean far to the north. The Techa was a vital lifeline for the communities of the region. Villages on the river and the surrounding floodplain depended on it for fishing, bathing, drinking and irrigation. The plant poured into it, on average, an Olympic swimming pool-sized volume of radioactive water every two hours. The official limit on the amount of radioactivity discharged in a day was 10 curies or less; in practice the plant often dumped ten thousand times this amount in a day. One way to avoid registering excessive discharge was simply to stop measuring. The plant's managers claimed that the secrecy around the plant prevented them from installing any instruments to measure discharge.

Some of the most radioactive waste liquid was produced when processing irradiated uranium fuel rods from the reactor, from which the plutonium was to be extracted. These rods were supposed to be stored in cooling ponds for six months before such processing, to allow some of the most potent and dangerous, but short-lived, radionuclides to decay into less "hot" nuclides. At Mayak this did not happen, and workers were forced to handle the rods – occasionally dropping and shattering them – with no regard for safety. Large quantities of water were needed in the extraction process used in the early days, known as the "all-acetate precipitation scheme", picking up high levels of toxic contamination, including high concentrations of sodium nitrate and sodium acetate, along with radioactive fission products such as cobalt-60, strontium-90 and caesium-137. Also present were valuable plutonium and uranium, which may have been one reason that it was decided that the waste should be kept, so that some of this material might be recovered at a later date.

Accordingly from around 1953 this fluid was fed into a series of colossal steel tanks, allowed to cool for a while, processed to remove some of the radionuclides, concentrated still further and returned to the tanks. The tanks were kept in a bunker concealing a "canyon" – a massive trench with 1.5 metre/5 foot-thick, stainless steel-clad, concrete walls, 8.2 metres (27 feet) deep. It was designed to hold 20 (although in practice there may have been only 16) giant stainless-steel containers, each with a volume of 300 cubic metres (80,000 US gallons), and each capped with a 1 metre/3 foot-thick, 160-tonne concrete lid. Even after reprocessing, the waste was still highly radioactive and thus generated a lot of heat. To keep them cool, the tanks were jacketed in water, with pumps to keep the coolant circulating around each tank. The whole trench was then backfilled with dirt. Soon after construction some of the monitoring instruments failed, but the intense radiation coming off the tanks meant they could not be repaired. In 1956, a cooling-water pipe leading to Tank 14 broke, but either engineers did not realize – because of the lack of monitoring – or it was simply too much trouble to fix, and so it too was left unrepaired.

Tank 14, insulated from the winter cold within its buried trench, and with no cooling mechanism, began to heat up. As the temperature inside reached 350°C (660°F), hot enough to melt lead, the water boiled away leaving a residue of acetates and ammonium nitrate; effectively the same as ANFO (ammonium nitrate-fuel oil), the high explosive used as a blasting agent, and favoured by terrorists such as those who bombed the World Trade Center in 1993 and Oklahoma City in 1995.

On the afternoon of Sunday, 29 September 1957, plant technicians noticed yellow smoke rising from the storage bunker. At 16:20, while they were still wondering what to do, there was a colossal explosion as the super-heated explosives in Tank 14 detonated with a force equivalent to over 70 tonnes of TNT. The immense concrete lid was hurled 25 metres (82 feet) clear. People miles away felt the impact of the 160-tonne projectile crashing to the ground. Around 75 tonnes of waste containing 20 million curies of radioactive material (nearly half of that released in the Chernobyl incident) was blasted a kilometre (1,095 yards) into the atmosphere, in a vast cloud of debris, smoke and glowing fragments. About 90 per cent of the radioactive material fell to the ground on or around the plant, but the rest was blown north-east in a long, narrow plume. A corridor of land approximately 300 kilometres (186 miles) long and 30–50 kilometres (18–30 miles) wide was dusted with strontium-90 at a density of at least 0.1 curies/km^2. Strontium-90 is particularly dangerous to human health because it "looks" like calcium in physiological terms, and so is readily incorporated into bones and teeth, where it lurks, firing out ionizing radiation. Strontium-90 exposure is thus associated with cancers of bone, bone marrow and soft tissues around the bones. Some 270,000 people lived in 217 towns and villages within this strip, which would eventually be labelled the "East Ural Radioactive Trace" or, in its Soviet acronym, VURS.

Management scrambled to formulate some sort of response. The plant director, Mikhail Demyanovich, could not be found; he was eventually located at the circus – in Moscow. Accordingly, not until Monday morning was the order given to evacuate the plant and immediate surroundings, during which time, more than 5,000 workers at the plant were subject to exposures as high as 100 rem. But even while the technicians and scientists were leaving, up to 20,000 prisoners and conscript soldiers were being trucked in to work as *likvidatory* – "liquidators" – , the name for Soviet clean-up workers, who were expected to "liquidate the consequences" of the accident. They had little protective gear or warning, let alone training, and only the most basic tools with which to clean up lethal contamination.

Over the next few days these liquidators followed the classic Soviet playbook for clearing up radioactive contamination: bury everything. Radioactive debris was buried in trenches, while larger fragments of the tank were dumped in a nearby swamp. In the process many of the liquidators were exposed to dangerous doses of radiation and suffered radiation poisoning; in the course of two years of clean-up operations, nearly 30,000 liquidators would receive doses higher than 25 rem. Far more people were exposed to damaging levels of radiation due to the ponderous response of the authorities in respect of protection of civilians. Not until more than a week had passed was there any effort to evacuate the local area.

In the immediate vicinity of the plant the forests were contaminated with blisteringly high-density radiation. In a 20 square-kilometre (8 square-mile) area where the contamination exceeded 180 curies/km^2, the pine needles received 3,000–4,000 rads in the first year, and all the pine trees died by the autumn of 1959. The three nearest villages in the affected zone were Berdyanish, Saltikovka and Galikaeva; their combined population was between 1,050 and 1,900. No one had explained to them what had happened or warned them about the lethal nature of the dark cloud that swept over them. In Saltikovka, for instance, at least one person, a 10-month-old girl, had died as a result; she had been in the garden with her mother when the fallout rained down, and died after suffering severe diarrhoea. Meanwhile in the village of Korabolka, according to one account, 300 of the 5,000 residents died of radiation poisoning within days.

A week or so later, Red Army troops arrived and ordered everyone in Saltikovka to evacuate. The villagers were forced to strip, dress in new clothes provided by the soldiers, and leave behind every single thing they owned. They were carted off on trucks as the soldiers razed their houses to the ground and shot their pets and livestock, to make sure that no one would be tempted to return. The same pattern was repeated at Berdyanish and Galikaeva. No explanation was forthcoming. The only mention of the incident in the local newspaper was a story about an unusual display of auroral "northern lights" a week earlier, which accounted for the "intense red light, sometimes crossing into pale pink and pale blue glow" seen in the sky the previous Sunday night.

Meanwhile hundreds of thousands of other people in the region were left to live on contaminated land and survive on contaminated water. Periodically the authorities would broaden the evacuated zone; further waves of evacuation occurred eight months later and again a year after that. About 20,000 hectares (50,000 acres) of agricultural land were deep ploughed to bury surface radioactivity deep in the soil. Over 10,000 villagers were eventually relocated, and a large tranche of the VURS zone was eventually fenced off and declared a nature reserve, a designation it still enjoys today. At least a quarter of a million people in the contaminated region would eventually be exposed to doses of radiation exceeding the permissible yearly limit – possibly as many as 475,000, according to some sources. However, poor health monitoring of the regional population and lack of records means that there is no reliable estimate of the excess death rate that resulted.

The priority at all times was not public safety but plutonium production. Plutonium processing restarted at the plant within two months of the accident, and just a few weeks after that there was another terrible accident at the plant (see page 138). There was no acknowledgement by the Soviet authorities of the occurrence, let alone the scale of the Kyshtym Disaster, although it was hard to conceal the vast extent of the VURS. Leo Tumerman, a Soviet biologist visiting the region in 1960, with no awareness of the existence of the Mayak plant, let alone the disaster, was unnerved to pass, on the main highway to Miassovo, a large sign warning: "DO NOT STOP FOR THE NEXT 30 KILOMETRES! DRIVE THROUGH AT MAXIMUM SPEED!" As he sped along the designated stretch, he could see on either side of the road a wasteland of bulldozed dirt, marked only by chimneys, shorn of the houses to which they should have been attached. When he arrived at the genetics seminar he was to attend, local participants filled him in on what they all called the "Kyshtym Disaster". It was so named, they explained, because the actual location of the accident was not on any map and did not officially exist; Kyshtym was the nearest town that was on the map. The villages along the highway had been evacuated and the houses all burned to the ground to prevent looters stealing contaminated goods. Only the stone chimneys had survived the fires.

Although the authorities ruthlessly suppressed news of the disaster, word spread among the scientific community, and the metaphorical fallout would have profound effects on Soviet science and nuclear policy. For one thing, it proved a turning point in the long struggle within Soviet biology to combat the pernicious influence of Trofim Lysenko. Lysenko was an agronomist whose malign effect on Soviet science is now regarded as a cautionary tale of the perverting influence of ideology and authoritarianism. Lysenko was little better than a fraud, but his rejection of what were deemed Western notions of genetics found favour with Stalin, and he became a tyrannical figure in the Soviet scientific community, pushing a bogus genetic doctrine known as Lysenkoism, which argued that interventions such as grafting and hybridization can alter genomes and force evolution in ways that contradict conventional genetics. It became illegal to counter this deluded pseudoscience, and scientists who dared to stand up for scientific integrity were imprisoned and even executed. Lysenko's influence initially waned under Khrushchev, but as the new premier in turn became more despotic, he rehabilitated Lysenko, touting the benefits of the "fruitful and revolutionary Lysenko biology".

Meanwhile the suppressed community of geneticists rallied around a concerted attempt to alert their peers in the nuclear science establishment to the dangerous genetic effects of radiation and radioactive pollution. What passed for common knowledge in Western science was underground propaganda in the USSR, and anti-Lysenko geneticists linked the cause of classical genetics to the need to study and control the dangers of radiation. This movement for radiation genetics won the favour of the father of the Soviet A-bomb, Igor Kurchatov, described by his peers as the "tsar of nuclear physicists", but Khrushchev remained bitterly opposed to it. The shocking failure of Soviet science to grapple with the biology of radiation exposure was brutally exposed in the aftermath of the Kyshtym Disaster. There was inadequate medical knowledge about radiation sickness or how to treat it, and Soviet scientists knew little about measuring doses or predicting their effects, in the short or long term. For instance, there was not a single laboratory in the entire country that could perform pathology tests to check for chromosomal aberrations, and there were no supplies of chemicals to protect against exposure to radionuclides, or bone marrow stock for transplants.

These shortcomings profoundly shocked the nuclear science community, and once they had sided with the geneticists, Khrushchev was forced to back down. Classical genetics was legalized. Empowered, and appalled by the airborne pollution resulting from the Kyshtym "dirty bomb", nuclear scientists now pushed Khrushchev further. In the words of dissident Soviet scientist Zhores Medvedev, "they were no longer just an obedient group of experts". Their insistent lobbying helped push the Soviet Union to sign an international agreement to end atmospheric tests of nuclear devices.

In a landmark exposé of Soviet scientific malpractice, published in *New Scientist* magazine in November 1976, Medvedev was the first to confirm what many in the West had suspected, but of which they had previously had only inklings. Outside of the USSR, the first known report of the incident was an obscure notice in a Danish newspaper, which reported in 1958 that "diplomatic sources" had provided intelligence of a radioactive explosion in the Urals. The article attracted little interest, although it must presumably have caught the eye of an unusual commentator, for in the June 1958 issue of *Cosmic Voice*, the journal of the Aetherius Society, came further news of the event. According to the article, on 18 April of that year, the occupants of a flying saucer had telepathically contacted the Society's founder, legendary flying saucer contactee George King. One of the beings, claiming to hail from Mars Sector 6, reported to King that:

> Owing to an atomic accident just recently in the USSR, a great amount of radioactivity in the shape of radioactive iodine, strontium 90, radioactive nitrogen and radioactive sodium have been released into the atmosphere of Terra.

King was a London taxi driver with esoteric interests who claimed that in 1954, while he was doing the washing-up, he heard a loud voice proclaiming: "Prepare yourself! You are to become the voice of Interplanetary Parliament!" Subsequently he experienced telepathic communications and visitations from a variety of entities, ranging from Jesus to interplanetary beings, and started one of the earliest UFO religions, the Aetherius Society. This was a quasi-Spiritualist organization founded on the belief that King and others could channel messages from interplanetary beings and flying saucer pilots, primarily concerning perils such as nuclear devastation and the prospect of salvation from them. One of King's regular correspondents was Master Aetherius from Venus, and according to the article in *Cosmic Voice*, this august personage had a lot to say about the Kyshtym incident.

First, Master Aetherius claimed that for a few weeks it would be harder for his cosmic communications to get through, "because of the foolish actions of USSR". Next he reflected on the Soviets' failure to acknowledge the accident, pointing out that "they have not yet declared to the world as a whole, exactly what happened . . . [nor] how many people were killed . . . [nor] that they were really frightened by the tremendous release of radioactive materials from this particular establishment during the accident."

King's extraterrestrial contacts claimed that the Interplanetary Parliament was expending immense energies to clean up the radioactivity, which might have been news to the thousands of Soviet peasants still drinking contaminated water. Apparently, however, things could have been much worse, for the Divine Intervention of Aetherius and pals had saved 17 million people from having been "forced to vacate their physical bodies".

On the one hand it is striking that a UFO contactee should appear to have access to esoteric intelligence of a top-secret incident behind the Iron Curtain, with a high enough degree of accuracy to specify radionuclides. On the other hand, King's ethereal interlocutors did not seem to be aware that they were a year late to the party, describing in the present tense events from the previous year, raising the suspicion that someone from the Aetherius Society had picked up on the Danish newspaper report, which made the same mistake.

How much did Western intelligence services know about the Kyshtym Disaster? The Mayak plant was the focus of extreme interest from the CIA and US military, and it was during pursuit of aerial reconnaissance of the site that, in 1960, one of America's high-flying U-2 spy planes was shot down. This was one of the most dramatic episodes of the Cold War, and the pilot, Gary Powers, was captured and subjected to a show trial before being swapped for a KGB spy. Eventually, through a combination of aerial reconnaissance, Soviet informants and curiosity among Western analysts about villages that were disappearing from Soviet maps of the region, the Americans were able to work out more or less what had happened. The US Government, however, had no desire to publicize a disaster that might sully the popular image of the nuclear establishment, and potentially complicate the development and deployment of their own nuclear assets. They particularly desired to avoid invidious comparisons with the safety record of the Hanford plant, on which Kyshtym was modelled. This thinking was so entrenched that, even when Medvedev exposed the truth about Kyshtym in 1976, Western authorities actively undermined his narrative by downplaying the severity of the incident. One response, for instance, dismissed his report as "pure science fiction", and of course it didn't help that Master Aetherius of Venus had got there first. Even *New Scientist* wryly acknowledged that it had been "Scooped by a UFO!".

The outlines of Medvedev's story, however, were soon confirmed by others, including Tumerman, the geneticist who had come across the alarming road signs in 1960, and who now lived in the West. But the details remained obscure. In 1989 secret files started to leak out of the crumbling Soviet state apparatus, and with the collapse of the USSR a couple of years later the truth finally emerged. The Kyshtym Disaster is now rated as a 6 on the seven-level INES scale (see page 374), or even a 7, according, for instance, to a 1993 report by the United Nations Scientific Committee on the Effects of Atomic Radiation. At any rate, it is almost certainly the third worst nuclear disaster in history after Chernobyl and Fukushima.

Meanwhile the Kayak plant was still very much in operation. Although it had stopped processing weapons-grade plutonium in 1987, it continued to process spent fuel rods, and to pump out pollution. As a result of decades of poor waste management and pollution control, levels of contamination in the area are shocking. In particular, Lake Karachay, one of the reservoirs used to collect radioactive effluent, had become the most radioactive body of water in the world, holding an estimated 120 million curies of radiation. When it dried out after an unusually hot summer in 1967, sediments from the dry bed were blown across roughly the same area as the contaminated zone from the Kyshtym accident, dumping another million curies of radioactive material on the region. In 1990, around the point where the effluent was discharged into the lake, the radiation exposure of someone standing on the shore was up to 600 rems/hour – enough to give a lethal exposure in an hour or less. When Dick Shaw of the British Geological Survey visited in 1998, he recalled, "We were taken in a lead-lined vehicle, because the radiation was so great." Only in 2015 was discharge into Karachay finally halted, and the lake was covered over with rocks and concrete. It is questionable whether the vast reservoir of contamination trapped by the crude containment will stay where it is, and likely that one day it will leak into the Techa river. Meanwhile, as a result of more than 40 years of dumping into Lake Karachay, radioactivity has seeped into the groundwater and migrated outwards for miles in every direction. The total volume of contaminated groundwater is estimated to be more than 4 million cubic metres (1 billion US gallons), containing in excess of 5,000 curies of fission-produced radionuclides with half-lives in the range of 30 years.

For the people living in the region, the consequences of this radioactive burden, accumulated from both the Kyshtym Disaster and the chronic pollution, are felt as a dramatically worsened health burden, in a region already suffering from chronic poverty and poor health. Perhaps the most shocking stories from the region are those of the settlements of Korabolka and Muslumovo. Korabolka was a village of 5,000 residents. Half of them were ethnic Soviets and half ethnic Tatars. In the wake of the accident and the passing of the toxic cloud, it is claimed, the Soviet half of the population was evacuated but the Tatars were left in place. Many of them believe this was deliberate: an experiment to test the long-term effects of radiation poisoning. Supposedly the cancer rate in the village, now known as Tatarskaya Korabolka, is five times that of comparable, uncontaminated villages.

A similar story is told of Muslumovo, which is not only blighted by an epidemic of health problems that seem likely to relate to radiation poisoning, but also appears to have been part of a long-running experiment into the effects of such contamination. Sited on the Techa river, but just beyond the boundaries of the Kyshtym fallout zone, Muslumovo was considered too big to evacuate. From 1962 it was the subject of a decades'-long survey by the Chelyabinsk branch of the Soviet Institute of Biophysics, known as FIB-4. FIB-4 doctors regularly gathered blood samples from the children of Muslumovo, and collected organs taken from autopsies of townspeople who died. They even collected stillborn babies, which were preserved in glass jars. Many of the specimens in this ghoulish compilation displayed congenital malformations.

The findings of the FIB-4 physicians led them to believe that hundreds of Muslumovo residents suffered from a condition that Soviet health physicists had learned to recognize from the very beginning of the Soviet nuclear programme: chronic radiation syndrom, a disease related to extended exposure to low doses of radioactive isotopes. But none of the town's residents were told of the findings, and nor were they warned of the ongoing health risks of relying on the polluted Techa for their water. In 1999, when a concerned local physician conducted her own survey, she found that 95 per cent of the townspeople had genetic disorders, 90 per cent of the children suffered from anaemia, fatigue or immune disorders, and only 7 per cent of the population did not suffer from a serious health complaint. It may be noteworthy that many inhabitants of Muslumovo are ethnic Tatars and Bashkir – minorities who have suffered tremendous discrimination.

Today the Mayak plant is still in operation, and still mired in controversy and concern about its record on safety and pollution. The long-forbidden zone of contaminated territory remains fenced off as a nature reserve, and, in a pattern that has been repeated around Chernobyl, nature seems to be flourishing there. It seems likely that for most organisms the health burden imposed by chronic high background radiation is trivial, a least in relation to other limiting variables, and when set against short lifespans and/or large populations. What the longer-term effects might be on the genetic fitness of populations remains to be seen. Tellingly, the only animals that seem to be badly affected by the radioactive contamination in the area are small ground-dwelling mammals, such as moles and voles, which are directly exposed at close quarters to the radionuclides in the soil. The long shadow of the Kyshtym Disaster remains on, or rather, *in* the land.

Chapter 4

SPOILT MILK: THE WINDSCALE FIRE, 1957

Tom Tuohy, Deputy Manager of the Windscale Works plutonium plant in Sellafield, north-west England, laboured up the steps, hauling himself onto the roof of the colossal nuclear reactor. This was the fourth time that evening he had made the long climb, 21 metres (70 feet) up from the ground. He cursed the plant designers for not installing a lift, but like everything at Windscale, the giant pile had been planned and built at breakneck speed in a haphazard process that often seemed to have been made up as it went along. Which was precisely what he was doing right now.

Standing on the concrete, Tuohy could feel the heat through the soles of his shoes. He walked quickly over to the east inner inspection porthole and put his hand on the cover. It was hot. This was concerning enough, but the real fear came when he opened the cover and looked down. In the course of normal operations, he should not have been able to see anything; the pile was primarily composed of a huge mass of graphite, a black crystalline form of carbon more familiar as the lead in pencils. When Tuohy had looked through the port at 18:45, the rear face of the pile had been glowing an ominous red. At 19:30 he had been able clearly to see red flames, and at 20:00 the flames had been a vivid yellow, as they burned off the salt in the sweat stains left by the construction workers who had assembled the pile in the late 1940s. Half an hour later and the flames were blue, meaning that the fire was now hot enough to ionize nitrogen. Tuohy realized that it had reached an intensity that could burn through concrete, and that the roof on which he was standing could collapse at any minute. Quite apart from the imminent danger to his life, Tuohy knew that this would expose to the open air a nuclear reactor in flames, spewing forth enough radioactivity to contaminate the region – perhaps even the whole of Britain – for centuries or millennia to come. It was time for desperate measures.

The Windscale plant was Britain's answer to the American X-10 reactor at Oak Ridge (and, if they had known it, to the Soviet Mayak plant near Kyshtym). Its genesis, operation and nemesis lay in the geopolitics of the postwar trans-Atlantic relationship between Britain and America. During the Second World War the British had been intimately involved in the Manhattan Project, from its inception as a British programme, to the roles of key personnel such as William Penney, a mathematician and physicist who had worked at Los Alamos. Penney had personally witnessed both the Trinity test and the bombing of Nagasaki (the latter from an observation plane nicknamed the Big Stink). After the war, however, the Americans abruptly changed tack and decided to shut the British out of their atomic clubhouse. The passage of the Atomic Energy or McMahon Act, in 1946, made it illegal to share information about the US nuclear programme with foreign powers – even with allies. British scientists were sent home and told not to contact their American colleagues.

From this point it became a priority of postwar British prime ministers, starting with Clement Atlee, to get back in with the Americans. This would require Britain to gain membership of the nuclear club through its own efforts, and so the British began a kind of mini-Manhattan Project of their own. William Penney, who was still working for the Americans as late as the Operation Crossroads tests of July 1946 (see page 35), was tapped to lead the programme, and he determined immediately that the best course would be to build an implosion-style plutonium device, similar to the Fat Man bomb and the bombs used in the subsequent Crossroads tests. This meant that the British would have to construct their own version of the X-10 pile at Oak Ridge: a nuclear reactor that converted uranium into plutonium. Without access to the technology, expertise and experience of the Americans, and with dramatically fewer resources, the British would have to do the best they could.

The X-10 reactor used graphite as a neutron moderator, with graphite blocks piled up into the form of a massive octagonal stack, through which ran numerous horizontal channels. Into these channels could be pushed small cylinders of uranium, sealed in aluminium canisters. Up and down the front of the pile moved a platform, on which workers could stand to push the canisters through with long poles. As the canisters sat in the pile, the controlled chain reaction would bathe their contents in neutrons, transmuting some of the uranium into plutonium. The workers simply poked the canisters through from the front, so that they fell out of the back of the pile into a pool of cooling water sunk into the floor. After cooling down, the canisters were retrieved and the irradiated contents processed to remove the precious plutonium.

The British decided to build two X-10-style piles. They wanted a remote location for the top-secret project and chose a disused munitions factory site at Sellafield on the Cumbrian coast, in the remote north-west of England. The factory had been built there because it was beyond the range of German bombers; now the site would be beyond the range of public awareness, although the nearest town, Seascale, soon became known locally as "Britain's brainiest town", thanks to the concentration of scientists and engineers living there. The plant itself would be known as Windscale.

A key element of any nuclear reactor is the cooling system; radioactive decay produces heat, while a nuclear chain reaction produces even more. The British knew that a water-cooled design would be problematic, as a huge supply of water would be needed, and there was the risk of a steam explosion if the circulation somehow failed. At Windscale, the only ready supply of water would be the sea, but saltwater is corrosive and so would be unsuitable. The British designers decided to keep things simple and follow the only example they knew; like the X-10, the Windscale reactors would be air-cooled. Huge blowers or fans would blast air over the piles to carry off heat, venting through 14 metre/46 foot-diameter exhaust stacks – aka chimneys – each comprising 50,800 tonnes of steel-reinforced concrete and standing nearly 125 metres (410 feet) high. For each reactor there was a blower building containing eight 2,300 horsepower main blowers, two auxiliary ones and four shutdown cooling fans.

Sat within 44 metre/144 foot-high hangars, each of the two piles comprised 50,000 machined graphite blocks, each 200 millimetres (8 inches) square and 800 millimetres (31½ inches) long, weighing 2,030 tonnes in total, arranged in an octagonal stack 15 metres (49 feet) high and 7.5 metres (24 feet 6 inches) thick. Drilled horizontally through this stack of graphite were 3,440 fuel channels, into each of which could be loaded 21 cartridges or slugs of uranium fuel, each 30 centimetres (12 inches) long, to give a total capacity of around 70,000 slugs of uranium, with a combined mass of 183 tonnes. In addition, there were 24 boron control rods, driven by motors, which could be moved in and out of the core to damp down the reaction. An additional 16 control rods were suspended vertically above the core to function as the scram (the emergency shutdown). The face of the core where the fuel cartridges were loaded was shielded with a 1.2 metre/3 foot 3 inch-thick concrete bioshield, lined at the back with 15 centimetres (6 inches) of steel. Holes drilled in the shield allowed access to the fuel channels; these holes were sealed with plugs. As with the X-10, the plutonium production process was basic but effective. The narrow aluminium-coated uranium slugs were manually pushed through the fuel channels, by workers with steel rods, until they fell out of the back of the pile, into a 90 metre/295 foot-long, water-filled channel. Again as with the X-10, rods were collected later, once cooled, and processed to retrieve the plutonium.

Work started at the Windscale site in 1947, even as the British nuclear research programme rushed to build a much smaller experimental reactor at the Atomic Energy Research Establishment (AERE) at Harwell in Oxfordshire, in order to learn fundamental principles. At Windscale an army of 5,300 personnel worked at pace to build a nuclear reactor plant in less than five years. Colossal amounts of concrete were poured – up to 268 cubic metres (9,464 cubic feet) a day. Yet critical aspects of the design were still being worked out as construction proceeded.

One such aspect was the need to filter the exhaust air. In 1947, Terence Price, a young scientist recently arrived at

Harwell, raised alarms about the risk of radioactive contamination should something go wrong at Windscale. The uranium fuel pellets were wrapped in skins of aluminium to prevent the uranium from oxidizing (i.e. burning), and to trap dangerous radioactive products that would form when the uranium was irradiated. But if the sealed cartridges were somehow to break open, or if the unthinkable happened and there should be a fire in the pile, the air-cooling system had the potential to pump huge quantities of radioactive contamination out of the chimneys and, depending on the prevailing winds, across the countryside. Price warned that the chimneys would have to be fitted with filters.

The immediate reaction to Price's warning was a breezy dismissal. "Don't be so silly, lad," he was told by Leonard Owen, chairman of the production pile design committee, "Two tonnes of air go up chimney every second. Can't filter that." Owen was referring to the gale-force draught that the massive fans would blow over the piles; a wind of 32 kilometres (20 miles) per hour designed to sweep away 100 megawatts of heat from each pile, allowing the piles to be run at the high energies needed in order to yield the 100 kilograms (220 pounds) of plutonium a year for which the British A-bomb programme was calling (a single bomb required just 5–10 kilograms [11–22 pounds] of plutonium, but the British were intent on building a nuclear arsenal). National security demands outweighed safety considerations, and construction of the great chimneys proceeded without filters.

Price, however, did not let the matter lie. He escalated his alert all the way to Sir John Cockcroft, director of the AERE and soon to become a Nobel laureate for his pioneering pre-war work on subatomic particles. Thanks to an agreement worked out in 1947/8, in the wake of the McMahon act, the US had agreed to be more open to sharing with the British information on atomic power, and as a result Cockcroft happened to be heading out to America to visit the Oak Ridge site. While there he was briefed on the perils of a fire and the very real risk of radioactive matter contaminating the cooling airflow over the pile. The Americans had fitted a filter that trapped particles larger than 0.001 millimetres (1 micron) in diameter. On his return to the UK in 1948, Cockcroft decreed that the Windscale reactors would indeed have filters.

There was only one problem. At least one of the two chimneys was already half-built: the base and over 20 metres (66 feet) of the brickwork of the Unit 1 chimney were already in place. The best place for filters was in the gallery between the reactor pile and the chimney, but these areas were already constructed. Cockcroft stood by his guns; the filters were non-negotiable. Accordingly they had to be fitted to the tops of the chimneys, in the shape of bulbous, 200-tonne assemblies of steel, bricks and concrete, giving the two enormous towers their unique and distinctive profiles. The strange appendages were known as "Cockcroft's Follies". The precise nature of the filter material would cause further challenges; the filter packs comprised fibreglass tissue coated with oil, to trap escaping particles. However, the fibreglass was fragile, and could not withstand the power of the cooling draught, which meant that the blower power would have to be reduced, which in turn limited the power at which the reactor could be run and thus the rate at which it could produce plutonium. Additionally, the filters trapped only particulate exhaust; gas, such as the radioactive fission product iodine-131 (when hot), would not be mopped up; if it were to leak from the uranium cartridges somehow, there would be no way to stop it escaping.

Design flaws in the piles themselves would be harder to fix, and cut off from American expertise in the construction of graphite reactors, the British designers found themselves prone to unwelcome surprises. In 1948 the American authorities did allow a contingent of scientists from Los Alamos to visit Harwell and dispense some advice, alerting the British to the research of the Manhattan Project's brilliant emigré scientist Eugene Wigner, into the response of graphite to a neutron-rich environment. Hungarian physicist Wigner had shown that the impact of neutrons affects the atomic structure of graphite, knocking atoms out of the crystal lattice such that they gather at interstices in the lattice. One of the effects of this realignment is to push apart atoms in the lattice, causing the crystal to expand and thus making the graphite swell: potentially disastrous for a carefully engineered stack of graphite bricks. Fortunately, the British learned about this "Wigner Growth" phenomenon relatively early in the design process, and further studies showed that British graphite had slightly different properties to American graphite. The designers calculated that their piles would be able to operate for up to 35 years before becoming unsustainably warped, but the episode should have sounded warning bells about the vulnerability of a reactor system that relentlessly bombards itself with high-energy particles.

Overcoming myriad challenges, the British successfully started up the first pile at Windscale in October 1950, and by January of the following year the first plutonium had been produced and processed. In June 1951, Pile No. 2 came on line and Windscale began to produce plutonium at full capacity. Within a year, however, problems began to emerge. In May 1952, Pile No. 2 experienced an unexplained temperature rise in part of the core. In September, Pile No. 1 experienced a similar phenomenon. Operators were able to bring down the temperatures with the cooling fans, but the problem was only explained when a second visiting American delegation revealed another of Wigner's findings about graphite.

He had shown that the carbon atoms displaced in the graphite crystal lattice are excited to a higher energy level than normal, so that neutron bombardment of graphite causes it to build up a store of potential energy. Known as Wigner energy, it can accumulate over time, with the high-energy atoms migrating to form clusters. If triggered with sufficient activation energy, these clusters can rapidly release their stored energy, leading to fast and substantial increases in temperature. The unexplained heating incidents had been releases of Wigner energy, triggered by the normal heating of the piles. They were harbingers of potentially catastrophic releases to come, for the longer the piles ran, the more Wigner energy they accumulated. At some point the dam might burst, causing a disastrous temperature rise sufficient to ignite the graphite pile. The only way to avoid this doomsday scenario would be periodically to trigger lesser energy releases deliberately, in a process known as annealing. This is where the core temperature is intentionally allowed to rise through fission activity, until Wigner energy release is triggered, and then the reactor is shut down and the cooling fans are used to drive off the excess heat. Through periodic annealing, it was hoped that Wigner energy could be released in controlled bursts, analogous to lowering the level of water building up behind a dam by regularly letting out bursts of water.

Unfortunately, the annealing process in a large pile would prove complex, since neither Wigner energy build-up nor its release were uniform. The pile operators had to juggle control rods and blower strength to produce heating in different parts of the reactor and then control the energy release. It was more of an art than a science. Complicating matters further, the monitoring technology used to guide the operation was hopelessly inadequate: temperatures in the core were measured by thermocouples inserted at various points in the pile, but there were only 66 of them. The first annealing operation took place in August 1953, and thereafter such operations were scheduled to take place every 30,000-megawatt days of operation of a pile. Annealing was a time-consuming process that required taking the reactor out of operation, so there was pressure to do it as infrequently as possible, but the longer the pile was allowed to run without relieving the pressure, the greater the reservoir of Wigner energy that would accumulate and the higher the risk of a catastrophic release. Between the first anneal and July 1957, eight such operations were performed.

After the first few, the process seemed routine enough to be performed with minimal supervision, even though no attempt had been made to regularize or codify the procedure; as the official report into the Windscale fire would later observe: "It appears that there is nothing in the nature of a Pile Operating Manual". In the US, checklists were becoming the standard operating procedure of the nuclear weapons programme, but in the UK there were no written instructions or checklists. This failure would later be heavily criticized, but it should not have taken an accident to unmask it. As nuclear safety veteran David Mosey remarked in his 1990 opus *Reactor Accidents: Institutional failure in the nuclear industry*, "Even allowing for the benefits of long hindsight, it seems a particularly conspicuous oversight to give charge of a demonstrably tricky and unreliable operation to an individual without providing anything in the way of formal written guidance aside from a memo of less than a hundred words."

Meanwhile the British A-bomb programme progressed. On 3 October 1952, plutonium from Windscale formed the heart of an implosion-style device that successfully detonated under water, just off the coast of Trimouille Island, Australia, with a yield of 25 kilotons. But even as Britain celebrated becoming the third nation member of the nuclear club, the Americans promptly moved the goal-posts. On 1 November, less than a month after the British fission bomb test, the Americans successfully detonated a "super-bomb", the first thermonuclear or hydrogen bomb, which unleashed the power of nuclear fusion to produce a yield of 11 megatons. Having delivered a successful fission device, Penney's team was now tasked with building a fusion bomb.

The Americans had leveraged their know-how to build a bomb using deuterium, a heavy isotope of hydrogen. But the British, working without American help, drew up their plans around tritium, an even heavier isotope. Tritium does not occur naturally; the British would have to make their own, through neutron irradiation of lithium isotopes in the Windscale piles. This meant creating a new style of cartridge for the piles: short rods of lithium-magnesium alloy, about 1.25 centimetres (½ inch) in diameter, sealed in aluminium cans. To maximize the yield from each can, the cartridges had the thinnest possible aluminium skins. Yet their contents were extremely dangerous: lithium-magnesium is pyrophoric, meaning it ignites spontaneously on exposure to air. What is more, the tritium production process involved lithium-6 nuclides in the can capturing neutrons and decaying to give helium and tritium nuclei. Thus, the process created helium gas, which pressurized the thin-skinned cans so that they would burst if the temperature reached 440°C (824°F).

By 1957, political pressure on the British nuclear weapon project reached fever pitch. At the start of the year, with Britain humiliated by the Suez Crisis, and with her international standing at a low ebb, the ambitious and energetic Harold Macmillan took over as Prime Minister. Macmillan felt keenly that he needed to restore British pride, explaining, "It is partly a question of keeping up with the Joneses. Countries which have played a great role in history must retain their dignity. The UK does not want to be just a clown, or a satellite". Accordingly he was determined to achieve a British H-bomb, and with an international test ban looming he felt that Britain was in a race against time. By May 1957, Windscale had produced enough tritium for Penney's team to put together the first H-bomb test device, Short Granite, but its detonation was a relative failure, achieving a yield of just 300 kilotons. The British were sure that a follow-up device would fare better, but they would need more tritium.

The Windscale piles would have to work harder, at higher power, for longer periods between annealing shutdowns. The annealing period was stretched to once every 40,000 megawatt days, despite the transparently hazardous nature of its design: a giant stack of flammable carbon, interspersed with tens of thousands of cartridges of flammable metal protected by thin aluminium skins, baking in the heat of radioactive fission, steadily accumulating potential energy, all while giant fans blew air across it. The pile resembled a pyre, built of matches and dusted with poison, built around a hot oven, and constantly ventilated by giant bellows.

The in-built risks of the Windscale reactors were exacerbated by a toxic brew of managerial arrogance, carelessness and secrecy. The operators of Windscale cared little for the environment or for safety. They ignored reports warning of degradation of materials in the reactors; were happy to suggest blocking local lakes or digging a mine in a nearby national park in order to bury waste; and regularly dumped contaminated water in the Irish Sea. Little thought was given to worker safety. Pam Eldred, the wife of Wally Eldred, a scientist at the plant who would later become a senior figure, recalled, "I could always tell when my husband had been irradiated . . . because . . . his hair was standing on end when he came home."

By this time Windscale was under the control of the UK Atomic Energy Authority (UKAEA), constituted in 1954, but by 1957 already a body in crisis, with chronic organizational failures and staff shortages. At Windscale, for instance, 52 out of 784 technical and management posts were vacant. As well as the earlier unscheduled annealing events of May and September 1952, there had been numerous other incidents, and radioactive particles were regularly found around the site, but when a Medical Research Council report on the contamination was shown to Macmillan, he merely professed to being greatly concerned and ordered that the report should be suppressed.

The annealing operation of July 1957 had not been entirely successful, as parts of the graphite core had not released their energy. By October it was storing up a suspected 80,000-megawatt days of Wigner energy, and so a ninth anneal was scheduled. Conditions were not ideal because Britain was in the grip of the "Asian flu" pandemic, and some key staff were afflicted, most notably H. Gethin Davey, the general manager of Windscale.

At 11:45 on Monday, 7 October, the annealing procedure began, and the pile operators very cautiously started to heat it by turning off the cooling blowers and withdrawing the control rods to allow fission to ramp up. Over seven hours they gradually withdrew enough control rods for the pile to reach criticality, and once the pile was running at low power they tried, through juggling the positions of the rods, to concentrate the fission reactions in the lower front region of the core, which previous annealing operations had not reached. The target temperature they sought to achieve was 250°C (482°F).

Just after midnight on Tuesday morning, readings from the thermocouples indicated that most of the graphite was relatively cool, at between 50° and 80°C (122° and 176°F), while a single thermocouple showed 210°C (410°F). So far, so routine, as annealing typically started from a single hotspot and spread out. The procedure seemed to be going to plan, and so the operators inserted the control rods to shut down fission and let Wigner energy release drive the annealing. The only wrinkle was that two of the fuel channel temperature monitoring devices had shown readings of 250°C (482°F), which seemed anomalous because, since only the graphite could anneal, only the graphite should be heating up. However, the Pile Physicist on duty, Ian Robertson, was not in the best condition to consider the problem, as he was stricken with flu. At 02:00 he went home to lie down.

Seven hours later Robertson was back, and the Pile Control engineers had bad news for him: they said that the pile seemed to be getting colder not hotter. Fingers were later pointed at the engineers for getting this wrong; the official report claimed that while some parts of the graphite were cooling, "a substantial number of the graphite thermocouple readings showed steady increases". To those in the control room, however, the general tendency seemed clear and it was decided to give the process a boost by pulling out some of the control rods to allow further heating by nuclear fission. This took until the evening, with the rods thrown in again at 17:00, and shortly afterwards Robertson, clearly suffering from his flu, was sent home again. But once more, anomalous readings had been observed in some of the uranium monitoring thermocouples; one had risen from 300° to 380°C (572° to 716°F) in just 15 minutes, at one point going up by 30°C (54°F) in a minute.

The next day, Wednesday, 9 October, saw a gradual decline in the concerning uranium temperature, and the desired-for general increase in the graphite temperature. Things seemed to be back on track, except that one graphite-monitoring thermocouple showed steady heating of a single channel, 20-53, which climbed from 255° to 405°C (491° to 761°F) by 22:00. By this time the Pile Physicist had ordered the flues between the pile and the chimney to be opened so that air would flow over the pile and cool it, and when this produced no effect he ordered the fans to be started up in brief bursts, at midnight, 02:15 and 05:10. Most of the pile responded, but the rise in the temperature of the anomalous channel 20-53 was checked only briefly, and by midday on the 10th it was up to 428°C (802°F). Meanwhile the temperature of the air venting out of the chimney was also climbing steadily, topping 85°C (185°F) on Thursday morning. Potentially most alarming was an abrupt rise in the level of radioactivity detected in the air flowing up the chimney; at 05:40 on the 10th it leapt to 6 curies, and by 16:30 it was at 30 curies.

By this time the Pile Manager, Ron Gausden, had been called in. Blasts from the fans had failed to cool the rising temperatures, but they had clearly blown contamination out of the chimney, because a radiation monitor on the roof of the nearby Meteorological Station was showing high activity. It was clear to Gausden that a uranium fuel cartridge had burst. At the rear of the pile were complex arrangements of air sampling devices known as Christmas trees, which were supposed to poke their beaks into the back ends of the fuel channels to detect affected cartridges. Gausden ordered the scanners to be deployed, but they would not move. This was not entirely unexpected; it was known from previous annealing operations that at high temperatures the Christmas trees would get jammed. A backup monitoring device showed, in the words of the official report, "a large positive reading for particulate activity". Gausden called Tom Hughes, the Works Manager for the piles. Indicative of the chronic understaffing of Windscale, Hughes was also the Acting Works Manager for the entire site *and* the manager in charge of the chemical group. His role as pile Works Manager was relatively recent, and he had never actually visited the pile. In the control room, Hughes found Huw Howells, the Health Physics Manager, who had come to find out why radioactive contamination was coming out of the chimney. They decided to call H. Gethin Davey, the General Manager of the plant, even though he too was stricken with flu. There had been "a bad burst" they told him; find it and clear it, Davey told them.

Accordingly a crew was dispatched to put on full protective gear, climb into the hoist used to lift crews up the face of the concrete bioshield, hoist themselves up to channel 20-53, open the loading port, and have a look. It was 16:30 on Thursday and the thermocouple reading for the channel was an alarming 450°C (842°F). The workmen pulled out the metal cylinder that plugged the loading port and peered inside to see a bright red glare. Davey telephoned his deputy, an ebullient northerner of Irish descent named Tom Tuohy, the nuclear scientist who had led the team that cast the first tiny ingot of plutonium from the piles. Tuohy was at home in the nearby village of Beckermet, about a mile from the plant, nursing his wife and family who were stricken with flu. "Come at once," Davey told him, "Pile No 1 is on fire." Tuohy knew the consequences of such a catastrophe: "I told my wife and kids to stay indoors," he recalled, "and keep all the windows closed." The next thing he did was to discard his radiation dosage monitoring badge; he didn't want anyone to tell him he had exceeded his permitted dose and would have to go home.

The official report on the incident, prepared by William Penney himself just a couple of weeks afterwards, concluded that the pile had been set alight at some point during the second nuclear heating of Tuesday morning. Penney admitted that it was impossible to know exactly what had happened, but that the most likely scenario was that excessive heating had caused the bursting of either a uranium cartridge or one of the lithium-magnesium ones, probably the former. Once this happened, a vicious cycle was set in motion. As the highly reactive metals leaked out of their split cases, they would combust, injecting more heat into the system, which would set off more annealing, release more Wigner energy, heat more cartridges and lead to further bursts. It was known that the thin aluminium alloy cans were liable to failure at temperatures above 427°C (800°F), while the graphite of the core would ignite if it reached 450°C (842°F). Yet the poor set-up of the monitoring equipment in the pile – what Penney called "the inadequacy of instrumentation for the safe and proper operation of a Wigner release" – had made it hard to get proper warning about the problems until it was too late. "Moreover," he went on to say, "in the condition of stagnant air which had necessarily been created within the pile, there was no means of detecting the smouldering uranium cartridges, which we believe to have been a key event in the development of the accident."

At first, with the blowers turned off, the smoke and ash from the burning cartridges lingered around the back of the pile, but as the fans were switched on, the conflagration spread and the temperature of the air coming off the pile increased, so radioactive particles began to rise up the chimney. The more the operators blasted the pile with the fans in order to cool it, the more air was fed to the fire. In particular, the activation at 13:45 on Thursday of the shutdown fans had stoked the fire so that by mid-afternoon what might have been a smouldering slow-burn confined to a single channel was now "a serious fire raging in the neighbourhood of the 20/53 group of channels".

The burning cartridges threatened to unleash a devastating cloud of radioactive contamination, with dangerous fission products from strontium to plutonium threatening to rain down on the surrounding region. The result of Sir John Cockcroft's determination was all that stood in their way, as his filter follies proved their worth. As his son Christopher later reflected, "Had the filters not been there I would think a considerable part of the Lake District and Cumbria would have been put out of bounds, at least for agricultural use and perhaps for people." Terence Price, who had been among the first to call for the filters, observed that "the word folly did not seem appropriate after the accident". It was estimated that, in the course of the fire, Cockcroft's Follies caught over 95 per cent of the radioactive dust created; the leak of radioactive gases, however, was a different matter.

The first order of business was to try to clear what were apparently burning fuel elements in the channels, but they were stuck and could not be moved. Gausden told the workers to try to create a firebreak around the affected channels by clearing out the surrounding ones, but this too proved extremely difficult. Many of the cartridges seemed to be stuck, presumably because they were swollen from the baking heat, and conditions at the work face were brutal. Dressed in full radiation suits, with intense heat radiating from the burning core, the workers tried to ram the fuel cartridges out of the back of the reactor using steel rods, and when these ran out (as they became contaminated and had to be discarded) they requisitioned scaffolding poles from a nearby construction site. Men could stand only a few hours of this at a time, and someone was sent to the local cinema to round up workers who were off-shift. The only thing that made conditions bearable was the positive air pressure supplied by the cooling fans blowing air across the pile, so that a cooling breeze came out of the loading ports as they laboured to poke poles through them.

The workers on the loading face were in a race against time. By 17:00 the graphite was burning fiercely, and temperature readings were going past 1,200°C (2,192°F). At this point Tuohy arrived; he was dismayed to see that the poles the workers pulled from the channels were glowing yellow with heat and dripping molten uranium. In Davey's office the scientists were discussing measures to quench the fire, including smothering it using non-flammable gas such as argon or carbon dioxide. All agreed that the one thing they did not want to risk was pouring water on the pile; at such high temperature there was a very serious risk that the water would immediately break down into oxygen and hydrogen, with the potential to cause an even more catastrophic hydrogen-oxygen explosion. It was also possible that replacing air with neutron-moderating water might cause a criticality hazard.

Tuohy decided he needed to get a better look at the pile and made the first long climb to the roof of the reactor hangar, to look through the inspection port. He repeated the arduous journey at regular intervals, directing the workers in what Penney later called "brute force efforts", which cleared several rows of channels of uranium fuel. Yet still the fire burned hotter, as Tuohy observed the flames changing colour from yellow to blue and estimated that at least 120 fuel channels were on fire; there was a fear that sustained temperatures of over 1,200°C (2,192°F) would trigger more intense Wigner energy releases and set the entire graphite pile ablaze. "At about this time," notes Penney, "the use of water was first considered."

Despite the imminent danger, British reserve was the order of the day for workers on the site. The initial thought of site foreman and senior Union official Cyril McManus was that the fire might secure his members a day off. Wally Eldred, a then relatively junior scientist who would go on to become head of laboratories at British Nuclear Fuels Ltd, says he was told to "carry on as normal", while chemist Marjorie Higham thought it best to keep her nose out of it: "Things did go wrong so you just didn't take any notice. The less you know about it the less you can tell anyone else." The pervasive secrecy surrounding the project, and the habitual deference to authority, guided the reactions of everyone associated with Windscale. "You kept quiet," reflected local woman Mary Johnson, who had been born on the farm that once occupied the site, "But you know you were scared stiff really. Those who were working there . . . didn't want to be seen against the thing."

At around midnight, Davey and Tuohy had agreed that, as a last resort, they would have to try to drown the flaming reactor. Getting water into the pile, however, would be difficult, because no one had ever dreamed of trying to do so. The plant fire brigade was called in and the process began of jury-rigging hoses to blast water through the loading ports. By 03:44 on Friday the 11th, hoses were ready to be attached when called upon, but the engineers and scientists were still hoping to avoid this colossal gamble. Although there had been no significant quantity of argon available, a tanker full of carbon dioxide had been sourced. At 04:30 it was emptied onto the pile but had no effect. The graphite was burning and the temperature kept climbing.

By this time Davey, incapacitated by the flu, had been sent home, and it fell to Tuohy to make the call. At 07:00 he decided to risk the water, but, mindful of the possibility of an explosion – which might have laid waste to the immediate environs of the pile, and elevated what was already a disaster into an apocalypse for much of the country – all plant workers were ordered to get under cover. In a very British twist, this in turn meant that they had to wait until the shift changed over, and the water was not turned on until 08:55. A message was sent to the chairman of the UK Atomic Energy Authority, Sir Edward Plowden, in London. Brimming with understatement and stiff upper lip, it ran:

> Windscale Pile No.1 found to be on fire . . . at 4:30 pm yesterday . . . Position has been held but fire still fierce. Emission has not been very serious and hope continue to hold this. Are now injecting water above fire and are watching results. Do not require help at present.

Tuohy directed the fire brigade chief to feed in water through two channels, less than a metre above the top of the fire. He remained at the pile as the water began to flow.

Union leader McManus, who had been a commando in the war, was impressed. "He was standing there putting water in and if things had gone wrong with the water – it had never been tried before on a reactor fire – if it had exploded, [the district] would have been finished, blown to smithereens," he recalled many years later. "It would have been like Chernobyl . . .".

Supposedly everyone except Tuohy and the firefighters had been ordered under cover at this point. However, Morlais Harris, a young scientist at Windscale, claims that he had been left in mortal peril. Tuohy had sent him up onto the roof to monitor temperature sensors, and there he stayed. "I was on top of the reactor throughout," he told science journalist Fred Pearce many years later. "They had forgotten me."

Fortunately for Harris, there was no explosion, and the fire chief was able to ramp up the flow of water from 1,100 litres (290 US gallons) per minute to 3,030 litres (800 US gallons) per minute. In the words of the official report, however, "No dramatic change resulted; at 09.56 flames were still feathering out of the back of the pile." Tuohy had a flash of inspiration. With the workers no longer sweating at the loading face, manually dislodging fuel, it was no longer necessary to keep the blowers on. At 10:10 he ordered the fans to be turned off, and the fire immediately began to subside. At noon Tuohy called Davey to report that the fire was out, although water was poured onto the pile for another 24 hours. The hoses were finally turned off at 15:10 on 12 October, and the pile was pronounced cold. As for poor Morlais Harris, "When they remembered and came to collect me at ten o'clock the next morning, the whole site was flooded with radioactive water that had poured out of the reactor."

The fire was out but the controversy was just beginning. The men from the UKAEA may have felt that the "emission has not been very serious", but not everyone would share in their complacency. A series of fortunate circumstances had limited the amount of radioactive material that actually escaped. Even though 10 tonnes of uranium had melted and 5 tonnes had burned, along with many of the lithium-magnesium cartridges, little of the uranium or the more deadly fission products actually left the pile, thanks to an oxide crust that had formed on the pile in the heat of the fire. Of the particulate contamination that did escape up the chimney, Cockcroft's filters had caught about 95 per cent, but the 5 per cent that they missed constituted a grave hazard, while the filters had not been designed to stop radioactive gases at all.

The work of the plant – and the primary products of the pile – were top secret, and so although some scientific papers published after the accident let slip that plutonium was among the contaminants, this was quickly hushed up. It was not until a major review of the case in 1984 that it was openly admitted that the leaked radionuclides included plutonium-210 and tritium (although the latter release was of minor radiological significance), alongside isotopes of caesium and strontium. Around the same time the government was forced to admit that another top-secret weapons isotope, polonium-210, used as a neutron source to help "jumpstart" nuclear weapons, had also leaked as a result of the fire.

The most significant release, however, was of iodine-131, which was emitted in gaseous form and thus was expected to pass through the filters unhindered. By a stroke of luck, this did not happen. One of the constituents of the lithium-magnesium cartridges was an oxide of bismuth, which was intended to transmute into polonium-210. The fire had burst the cartridges and sent bismuth oxide dust up the chimneys where, alongside vaporized lead from some of the other canisters, it coated the glass fibres and reacted with the iodine gas, capturing it in the filters. Of around 70,000 curies of radioactive iodine present in the pile, 50,000 were released up the chimney but only 20,000 curies made it past the filters.

Thus at least 20,000 curies of radioactive material escaped as a result of the fire, and although this was a thousand times less than would later be released in the Chernobyl disaster, it was nonetheless enough for Windscale to classify, in the eyes of the West, as the worst nuclear disaster that had ever occurred. The Kyshtym disaster had occurred just one month earlier (see page 63), but it would not be widely known about until over three decades later, and until the Chernobyl disaster, Windscale was the worst incident of which the world was aware. The fire in Pile No. 1 has traditionally been rated as a level 5 incident on the INES scale, but reassessment of the accident, in light of the acknowledgement that plutonium escaped, has led to suggestions that it should be raised to a level 6: "Serious Accident".

The initial plume of radiation travelled north-east, but then a change in the wind direction carried radioactive contamination over southern Cumbria and the neighbouring county of Lancashire. Over the following days, the cloud of hazardous material drifted across the whole of north-west England, and as the iodine cooled it precipitated out of the atmosphere as radioactive dust, dropping as far away as Preston, 85 kilometres (53 miles) to the south, and Leeds, 140 kilometres (87 miles) to the south-east. Eventually the cloud crossed the English Channel and dropped contamination on Holland, Germany and even Norway. One of the unresolved controversies around the fire was the lack of warning given to those living in the local area, let alone further afield.

Shortly after midnight, very early on the 11th, when the plant managers started seriously to consider risking an explosion by using water to douse the fire, the chief constable of the region had been warned of "the possibility of an emergency", as Penney described it. A critical observer might argue that the fire had long passed that milestone; why was he not warned about the radiation already leaking out of the chimneys? Penney himself addressed this issue in his report:

> It was represented to us that warning of an emergency ought to have been conveyed to the inhabitants of the area surrounding the Works. Two witnesses reported that high levels of activity had been measured on grass and on clothing in Seascale [the nearby town], and on clothing of people cycling to work along the track from Seascale on the morning of 11th October. The Health Physics Manager was satisfied, from the district measurements already mentioned, that no district radiation or inhalation hazard existed: there was therefore no occasion to issue a district emergency warning, which would have caused unnecessary alarm.

Penney goes on to claim, effectively, that the two witnesses were making a fuss about nothing, since the activity measured on their clothes was "some 20 times lower than that which would have constituted any hazard in accordance with the standards observed by the Authority and based on the Medical Research Council tolerance levels". But some of those living in the shadow of the plant already knew the hazard was real. In Seascale, Morlais Harris's father pointed a Geiger counter at his lawn; it went haywire. Plant workers sent word to their families to leave the area fast, yet other locals carried on unaware. Jenny Jones, then a young girl attending Calder Girls' School, less than a mile from the burning pile, recalled in a recording made many years later for an oral history of Windscale that "when I went home that evening, my mum and dad knew something had happened . . . they'd noticed that the village had gone very quiet. Then we got a phone call from one of the daily papers, asking if there was any panic in the village."

Despite Penney's characterization, the Health Physics Manager, Huw Howells, was far from satisfied that there was no risk to the public. He knew that iodine-131 posed a potentially terrible hazard, specifically to children and pregnant women. When humans consume iodine it is rapidly concentrated in the thyroid gland, and such a confined mass of radioactive iodine, emitting beta and gamma radiation, dramatically raises the chances of thyroid cancer, particularly in children. What scared Howells was that the land around Windscale was dairy farming country, and iodine-131 that was raining down on pasture would contaminate grass, which would be eaten by cattle, and pass straight into their milk. There was very little science on what constituted a dangerous level of iodine-131 contamination of milk, but Howells was aware of one paper that suggested anything above 0.39 microcuries (μc) per litre (2 US pints) would be hazardous to infants. Milk samples were urgently gathered from nearby dairies. An initial analysis showed peak levels of 0.48 μc/l, but milk gathered at Seascale on the morning of Saturday, 12 October, showed levels of 0.8 μc/l. Later Howells and his team would set an exposure limit of 20 rads for any individual child, which in turn suggested an iodine-131 limit of just 0.1 μc/l. In other words, the milk was dangerous. Howells told Davey that they needed to stop the distribution of milk.

Davey called in the local police and the Milk Marketing Board (MMB), and that night constables and MMB workers roused farmers from their beds to warn them not to send out the milk from their dairies. Instead they would have to pour it down drains and into ditches, and it was said that the countryside smelt of spoilt milk for some time afterwards. The Atomic Energy Authority offered to compensate farmers for their lost milk, but the process soon descended into acrimony. More widespread sampling of land and milk suggested that contamination had spread much further than realized. Site foreman Cyril McManus, one of those dispatched to gather samples from as far away as Devon, hundreds of miles to the south, said of the region around Windscale, "there was contamination everywhere, on the golf course, in the milk, in chickens". On 15 October, with contamination detected as far afield as North Wales and Southern Scotland, the ban on milk distribution was extended to over 500 square kilometres (190 square miles), encompassing over 600 farms.

All the while, however, the authorities were trying to keep a lid on the whole affair. With farmers becoming angry and suspicious, journalists prying and the public increasingly alarmed, the aftermath of the fire was fast becoming a public relations battle, one in which cows would prove to be the unlikely foot soldiers. In the days after the accident, health officials gave assurances that livestock would be unaffected, and a dairy cow from a farm near Windscale was taken to the Manchester studio of the regional television station, Granada TV, and milked on air, with its milk then tested with a Geiger counter. Many in the farming community were not mollified; rumours circulated that cattle poisoned by radiation had been seized by the authorities and slaughtered. Davey arranged local meetings to address the farming community directly. At a meeting in Gosforth Public Hall on 24 October 1957, farmers affected by the milk ban told Davey they felt as if they were "regarded almost like lepers by their colleagues outside the affected area". William S. Newall, agricultural correspondent for local paper *The Whitehaven News*, wrote that the meeting was so tense that it was as if the representatives of Windscale were "being thrown to the lions with a vengeance". Some of the farmers, Newall reported, lamented that their cows were "doomed and their farms valueless" and they were determined to "seize on any and every rumour to add fuel to their fires of resentment". Problems were compounded by the uneven length of the ban on milk sales. For most of the farmers, the ban was lifted on 29 October, but for some it continued until 23 November. Over 3 million litres (790,000 US gallons) of milk was destroyed.

Spoilt milk caused problems for the nuclear authorities beyond the dairy farmers. According to Richard Batten of Exeter University, a historian of British nuclear power, "news of the milk ban created more anxiety within the general public than the fire itself". Fear of contamination turned public opinion against not only local dairy farmers but also other farmers and even fisheries. In a counterpunch to the government's TV cow, a cow from Cumbria was taken to the Granada TV studio and broadcast on national TV with the label "Radioactive Cow". Cattle from the affected area that went on sale at the Whitehaven cattle market the week of the fire were marked with a special yellow sign, resulting in what one dealer called "buyer resistance". Two farmers from Devon who had bought cows from a Cumbrian market were, it was reported, ordered by their local police force to destroy the milk that was produced. Anxiety about contamination was liable to spread to any related industry. In February 1958, chocolate manufacturer Rowntree's filed a compensation claim with the AEA, claiming that they had been forced to dispose of large quantities of Chocolate Crumb because the processed milk among its ingredients threatened to damage their public image. The AEA eventually agreed to pay for the milk used by the company to be destroyed. "We would like to do this as discreetly as possible," noted Rowntree's, "so that the name of this Company does not become associated with radioactive contamination".

Some farmers believed that the fallout had affected the fertility of their livestock. George Curwen, a Cumbrian farmer and local leader of the National Farmers' Union (NFU), said that he had received dozens of complaints, and that there was "no doubt about it that the fall-out is to blame". Curwen distrusted the commitment of the AEA to compensating farmers: "Give them time enough and they'll dodge the issue altogether". Over time, however, opinions mellowed. A year after the fire a Captain Craven Hodgson claimed that although half his herd had shed large patches of hair, they were calving better than they ever had. An article in *The Whitehaven News* from 25 December 1958, trumpeted two "Winning 'Windscale' Bullocks" as "confirmation of findings by the NFU and the Ministry of Agriculture that nuclear operations on the West Cumberland coast . . . [had] . . . no ill effects on local dairy cattle and beef industries", accompanied by a photograph of Hodgson and his wife, with their prize-winning animals, in front of the Windscale plant.

So what was the real story about the true danger to the public, and the actual harm that resulted from contamination from the fire? The official position, to begin with, was that "there has been no immediate damage to the health of any of the public or of the workers at Windscale, and it is most unlikely that any harmful effects will develop". This, at least, was the verdict of Penney, though since it was given less than a month after the fire, it must be judged a little complacent. Nonetheless, Penney was supported by the UK Medical Research Council, which similarly downplayed any wider or longer-term risk to the public. Over the following decades, however, a slightly different picture emerged.

In 1979, for instance, researchers at Manchester University found elevated incidence of mortality from leukaemia in the cities of Lancaster, Preston, Blackpool and Burnley – all relatively close to Windscale. They were not able, however, to prove a connection with the Windscale fire contamination. Subsequent studies based on more nuanced understanding of the effects of radiation exposure on cancer risk, estimated how many deaths are likely to have occurred, given the known radiation release. One estimate for excess deaths in England up to 1990, attributed around 100 fatal cancers to the fire; mainly lung cancers as a result of inhalation of polonium-210, one of the radionuclides that the government had denied was even present. Alongside these were around 90 non-fatal cancers, mainly due to thyroid cancers resulting from iodine-131 exposure.

Other estimates have suggested around 240 thyroid cancers resulted from the fire, while the authoritative history of the disaster, Lorna Arnold's 1992 book *Windscale 1957: Anatomy of a Nuclear Accident*, claims that up to 248 cancer deaths may have resulted from the fire. While death tolls in the hundreds sound alarming, and each individual case is a personal tragedy, they are statistically insignificant in the context of overall death rates, making it hard to counter definitively the claims of the authorities at the time.

Similarly, the health and excess mortality effects on those directly involved in the accident have been found to be statistically insignificant, and thus effectively undetectable. The 471 workers who were active in fighting the fire or in clean-up operations received relatively low mean external radiation doses of 0.6 rem in October 1957; any effect of this would be swamped by the average lifetime occupational exposure of such workers, which was 32 rem. Much bigger and more physiologically problematic doses might have occurred as a result of inhalation of particles, but there is no data on this. A 50-year follow-up study of the 471 workers found no evidence of an influence of the fire on their health, nor any significant difference in the cancer rate between them and other workers at Windscale or Sellafield (as the plant was later renamed). Tom Tuohy, for instance, who was at the heart of the fire-fighting operation and repeatedly looked through the inspection port directly at the burning reactor pile, died in March 2008 at the age of 90.

While the health burden resulting from the fire seems, thankfully, to have been low, its effects were dramatic in terms of public opinion and the cultural and political discourse around nuclear energy. Even the prospect of this greatly alarmed the British establishment. The Windscale fire was almost certainly ultimately attributable to the pressure piled on the UKAEA to produce material for Britain's nuclear weapons programme. The demand for plutonium, tritium and other isotopes drove the plant operators to trim ever finer the margin of safety, while stuffing the pile full of canisters and running it for longer periods between shutdowns. The aim of the entire exercise was to produce a thermonuclear weapon, and on 8 November, less than a month after the fire, it succeeded. The British detonated the H-bomb Grapple X near Christmas Island in the Pacific, achieving a yield of 1.8 megatons. Macmillan had been gifted a valuable card to play in the delicate negotiations with the Americans with which he was engaged. Desperate to avoid jeopardizing these talks, Macmillan promptly buried the Penney report into the fire. Despite finding it "fascinating . . . [and] prepared with scrupulous honesty and even ruthlessness", the Prime Minister blocked its publication.

Instead the government published a much less damning version of the report, with a foreword written by Macmillan himself, in which the accident was attributed partly to shortcomings in the monitoring system of the pile, and partly to poor judgment by unnamed personnel. This sparked a media witch-hunt for the culprits, and the finger of suspicion was levelled primarily at Ron Gausden, the pile group manager, and his colleagues. It was a deeply unfair outcome, which the government allowed – and in effect tacitly encouraged – by covering up the real, underlying causes of the accident: poor design, underfunded management and excessive demands from the nuclear programme. Gausden left the UKAEA in 1960 to join the newly created Nuclear Installations Inspectorate, where he went on to a enjoy a distinguished career.

As for the pile itself, the future looked bleak. A contemporary report on the fire in the science journal *New Scientist* pointed out that:

> The engineers have not only to clear the burnt-out channels of radioactive debris, but to dry out the pile in a way that will remove all the waterborne chemicals, such as calcium, which would interfere with the fission process. Unless these products are removed, the future efficiency of the reactor will be impaired; but it has to be seen whether this can be done without going to the lengths of stripping down the entire reactor and rebuilding it.

In fact, neither Pile No. 1, nor Pile No. 2 which had been shut down five days after the fire as a precaution, would ever be used again. They were dinosaurs even before the accident; their replacements, the four Calder Hall Magnox reactors, were already in place at the plant, and had been operational since August 1957. These reactors generated electricity, making them the world's first civil nuclear power stations, although they also, more importantly for the government, produced plutonium and tritium. Air-cooled piles, like No. 1 and No. 2, would never again be constructed; Magnox reactors used closed-loop carbon-dioxide cooling systems.

Instead the two reactors were, as far as possible, dismantled. As much uranium as possible was recovered from Pile No. 1, some radioactive debris was dumped in the sea and the cores of both piles were sealed off with concrete. Yet within Pile No. 1 there remained 6,700 damaged fuel cartridges – up to 15 tonnes of uranium – and 1,700 special isotope cartridges. Decommissioning did not begin until 1988 and is not expected to finish until the 2040s. Even the chimney for Pile No. 1 was too contaminated to demolish for decades. When work did start on pulling it down, it proved to be extremely complicated, and was halted altogether after the death of a worker. Not until 2018 was the structure finally demolished, with the help of a specially constructed 152 metre (500 foot) crane. The son and grandson of Sir John Cockcroft looked on as his iconic filter galleries finally came down.

The shadow cast by the Windscale fire has loomed for almost as long. The accident helped to ignite antinuclear sentiment that had, until then, been merely smouldering. In 1969, Glenn Seaborg, Nobel prize-winning American nuclear chemist and latterly chairman of the Atomic Energy Commission, battling a rising tide of antinuclear activism, lamented of critics that "they always go back to Windscale". The Windscale brand was so tarnished that in the 1970s the site was renamed Sellafield. Unfortunately, this name, too, has become synonymous with bad press and public suspicion around radioactive pollution and leaks, as the plant was repurposed as a nuclear waste processing and storage facility, which suffered a series of damaging incidents. For instance, the Thorp reprocessing facility at the site had to be shut down in April 2005, after a broken pipe spilled acid in which was dissolved 20 tonnes of uranium and 160 kilograms (353 pounds) of plutonium. Sellafield is now stuck with vast amounts of radioactive material and waste, including 140 tonnes of plutonium and enough high and intermediate level radioactive waste to fill 27 Olympic-sized swimming pools. Some of the waste will remain radioactive for 100,000 years. It costs UK taxpayers billions of pounds a year to run the site and the clean-up may continue for another century.

One of the strangest aspects of the Windscale fire is its proximity in time to the Kyshtym disaster. In a 2017 article in the *Journal of Radiological Protection*, "A double diamond anniversary—Kyshtym and Windscale: the nuclear accidents of 1957", Professor Richard Wakeford of Manchester University noted, "That the first two nuclear accidents having serious off-site consequences happened within two weeks of each other seems a rather remarkable coincidence, particularly when the next such accident—the Chernobyl accident in 1986—occurred nearly 30 years later." He went on to consider whether the coincidence is mere "fluke" or points "to some underlying commonality?". Comparing the two incidents reveals plenty of commonalities, from underlying agendas (pursuit of accelerated nuclear weapons programmes), to problems with build and operation (hasty construction and insufficient safety protocols), to the plants outliving their originally intended lifespans (both were 10 years old when the incidents occurred). In both cases, Wakefield concluded, "accidents were just waiting to happen".

Chapter 5

BURNING UP: THE GRIM DEATH OF CECIL KELLEY, 1958

THE ONLY thing more dangerous than tickling the dragon's tail is doing it without knowing. Critical excursion victims such as Harry Daghlian and Louis Slotin were at least aware of the perilous game they played. What makes materials such as plutonium so treacherous is that the margin between sub- and supercritical can be both infinitesimal and mutable. Despite the popular coining, criticality is a function of much more than mass; there are multiple parameters that govern its boundaries, and they can change. When someone is working with fissile material in ignorance of their proximity to the threshold of criticality, and when that threshold can literally change at the push of a button, the ingredients are in place for a terrible accident. On 30 December 1958, at Los Alamos, experienced chemical engineer Cecil Kelley was the one tickling the dragon's tail without even realizing it, and he was about to get burned.

Los Alamos had been through some rough times after the war, but as America relentlessly ramped up its nuclear weapons programme, it would become one of the central nodes of the military-industrial complex. It grew into a multi-billion-dollar facility, pursuing multiple strands of research, while continuing with its core mission to build nuclear bombs. This included aspects of chemical engineering such as the dirty, difficult and dangerous task of recovering plutonium that might otherwise go to waste, through chemical separation methods. The workshops at Los Alamos generated waste plutonium from processes such as casting and machining, as the precious metal was moulded and shaved into bomb cores and the like, and it was both dangerous and inefficient to dispose of that waste. Accordingly it had to be recovered, and the first step was to dissolve it in large quantities of solvents. The primary solvent was nitric acid, an aqueous (water-based) solution, but organic solvents (essentially types of oil) were also used. The recovery process involved multiple large tanks, with solutions piped and poured from one to another as they were concentrated and purified, mixed and separated and mixed again.

The tanks in question were housed in a laboratory known as DP West, in a windowless concrete room. They were stainless steel cylinders around a metre in diameter, with a capacity of 1,000 litres (264 US gallons), standing on squat metal legs. They had rounded tops and bottoms; two small glass portholes at the top allowed the operator a view of the contents, while sight gauges up one side made it possible to monitor the level of liquid. At the bottom of the tank was a rotating impeller for mixing the contents. At the push of a button the blades took just a second to accelerate to 60 revolutions per second.

Unfortunately the design of these tanks was inappropriate for the dangerous task assigned to them. Their rounded, barrel-like nature greatly decreased their surface area to volume ratio. Fissile material in solution is still fissile and capable of going critical, with the correct geometry. If the geometry of the solution means that enough fissile atoms are close enough together, criticality – possibly supercriticality – results. After seven years of using these tanks on a daily basis, Los Alamos had finally woken up to their design flaws; about a month earlier, the laboratory's Nuclear Criticality Safety Committee had reviewed the entire plutonium process and recommended changes. Richard Money, a chemist at Los Alamos whose career overlapped with Kelley's, recalled that, "By that time, apparently, the laboratory . . . had determined safe geometry tanks for solutions". Short, fat tanks were the wrong shape; tall, narrow ones made much more sense. If the diameter of the tank were small enough, then no matter how concentrated the solution it contained or how tall the tank, it would be impossible for a critical volume to be achieved. As Money put it, "You could go up to heaven with it and fill it up with plutonium or U-235, and because of the geometry, it would not be dangerous". The Safety

Committee had called for the tanks to be replaced with pipe sections that were each 3 metres (10 feet) long but just 15 centimetres (6 inches) across.

Accordingly, the tanks at the DP West site were due to be replaced in the summer of 1959, but in the meantime it was assumed that work practices which had avoided mishaps for the previous seven years could be safely followed. In 1958 the man in charge of the tanks was Cecil Kelley. Interviewed in 2009, Richard Money recalled him as a "wonderful, wonderful guy . . . He never went by the name of Cecil, he went by the name of Kelley. Even his wife called him Kelley. Everybody knew him by Kelley, 'Hey, Kelley.'" Kelley was working the last shift before the facility closed for the New Year's holiday. After eleven and a half years at Los Alamos, he must have assumed his task that afternoon would be routine, but in fact the lab was in the middle of a non-routine operation: an annual physical inventory, in which the lab sought to evaluate the plutonium content of the residues left in the various process vessels (i.e. tanks and such). Or, as Money put it, "Coming toward the end of the year . . . all the groups are trying to get rid of their waste to get them off the books." As a result, Kelley was "getting a little bit overloaded with material to process". Unfortunately, he did not realize just how overloaded. What no one had appreciated was just how much plutonium was in those residues, because there had been a failure to perform the analyses necessary to keep rigorous tabs on every gram of plutonium, according to Money. Thus Kelley had no idea how much had been washed out of supposedly empty tanks to converge in his vessel. "Unbeknownst to him," recalled Money, "a sludge of fissionable material . . . was building up in . . . his tank".

Raised up on its stubby legs, the tank was taller than Kelley. He stepped up onto a low foot ladder in order to peer in through the porthole in the lid of the tank, at what he believed to be a combination of "lean" solutions. Kelley was expecting the solution in the tank to contain less than 0.1 gram of plutonium per litre (0.75 grains per pint), but in fact the concentration was 200 times higher than this. Even worse, the plutonium content was unevenly divided between two layers of fluid in the tank. The lower layer was 300 litres (79 US gallons) of aqueous solution containing a total of 60 grams (2 ounces) of plutonium, while the upper, much more concentrated layer comprised 160 litres (42 US gallons) of organic solvent containing 3.1 kilograms (6 pounds 11 ounces) of plutonium; enough to go supercritical if rearranged just so. Increasing the risk of criticality, the hydrocarbon nature of the organic solvent made for an excellent neutron moderator, while the aqueous layer beneath had the potential to act as a reflector, bouncing escaping neutrons back towards the upper layer. Yet this oil-and-water combination was expected; in fact it was precisely the purpose of the tank to mix aqueous and organic fluids together to give a homogenous emulsion, from which the plutonium could then be recovered. Normally, there would be nowhere near enough plutonium atoms in either the mixed or unmixed fluids to constitute a criticality risk. Accordingly, at 16:35, Kelley pressed the switch to turn on the impeller.

The impeller immediately started to rotate the mass of water that comprised the bottom layer, but inertia meant that the thin layer of oily solvent on top moved much more slowly. The aqueous layer assumed a whirlpool shape, rising up around the sides and getting lower in the centre to create a very shallow bowl, into which slumped the oily, plutonium-rich upper layer. The thickness of this layer, according to a contemporary press report put out by United Press International (UPI), and based on a detailed briefing by the Atomic Energy Commission, "was increased [by] an average of two-fifths of an inch [~0.5 cm]. It was enough to change its shape into a 'supercritical configuration' . . . It is literally true that a plus of two-fifths of an inch killed Kelley". This tiny change in the geometry of the plutonium-bearing solution brought enough plutonium atoms close enough together for just long enough to constitute a critical assembly, capable of going prompt or supercritical.

This minute alteration in the thickness of the oily layer occurred within a second of Kelley starting the impeller. The plutonium in the solvent went supercritical for 0.2 seconds, during which time 150 million billion plutonium nuclei split apart, releasing a massive blast of neutrons and gamma rays. Kelley felt a wave of intense heat, and either fell or was knocked off his ladder, as the tank lurched a centimetre (½ inch) sideways. A pair of workers on the other side of the room noted a flash of light cast onto the ceiling, variously described as a blue flash of ionization light or a spark from a short-circuit in the impeller motor mechanism. They also heard a dull thud, which may have been the tank shifting.

Kelley got to his feet, disoriented, and turned off the impeller. Then he turned it on again. Richard Money believed that Kelley had thought to himself, "'Of course, it overloaded the circuit breaker and it knocked off the stir, so the stir stopped.' He thought that a circuit breaker, the circuit breaker had gone off, so he goes over to the wall and he flips the switch again." Money surmised that this caused a second criticality event, but in fact bubbles produced by the energy released in the excursion, together with the mixing achieved by the impeller, dispersed the plutonium throughout the fluid sufficiently that the concentration dipped below 7 grams per litre (⅛ ounce per pint), too low to go critical again.

Kelley stumbled outside into the snow, followed by his two co-workers. They found him ataxic – experiencing a loss of co-ordination – and mumbling, "I'm burning up, I'm burning up". This, pointed out Joseph L. Myler, the journalist who wrote the contemporary UPI report, "was significant. You can take killing amounts of radiation . . . without feeling a thing." What Myler meant is that a dose of radiation great enough to be lethal can be imperceptible;

the fact that Kelley felt he was burning up showed just how massive a dose he had received. Indeed, as Money later recalled, "It was far more radiation than anybody else has ever received in peacetime."

None of this was apparent at the time; radiation alarms had not been triggered and the two co-workers who had come to Kelley's assistance had no clue what had just happened. They thought perhaps some chemicals had spilled on him, and half-carried Kelley to the emergency shower, turning off the tank stirrer as they went past. Emergency responders soon arrived on the scene, including a nurse, supervisors and radiation monitoring staff. Kelley was in shock, barely conscious, and his skin was reddening fast, prompting the nurse to exclaim that he had "a nice pink skin". She did not realize that in fact he was exhibiting acute erythema as a result of massive radiation exposure, similar to a severe sunburn. Except in Kelley's case, while his skin was the only visible evidence, every layer of tissue in his body, front to back, had been burnt.

The nurse was not the only one to fail to appreciate what had occurred; the radiation monitors checked the tank area with alpha detectors – devices that pick up alpha radiation, which might reveal the presence of leaked or splashed plutonium. Since this was not the cause of the incident, they did not find anything. It was only a few minutes later, after Kelley had been loaded into an ambulance and was being driven away, that they started doing gamma radiation measurements. When they recorded rates of tens of rads per hour around the mixing tank, the monitors realized that an excursion had occurred.

When Kelley arrived at the Los Alamos Medical Center he was entering the second stage of very severe acute radiation syndrome, vomiting, hyperventilating and shaking uncontrollably, so that his limbs had to be restrained to allow the medics to work on him. His skin was, according to a 1995 Los Alamos review of the incident, "dusky-reddish violet" but cold, and his lips were blue, indicative of low oxygen levels in his blood. Nurses were unable to obtain his pulse or blood pressure. He was wrapped in blankets and surrounded with hot water bottles, but it was not until he was plied with sedatives that his shaking calmed down, at which point his pulse was measured at 160 beats per minute and his blood pressure at 80/40.

An hour and forty minutes after the incident, Kelley emerged from his shock reaction and became relatively coherent and comfortable, although he complained of abdominal pains and occasionally retched. His body, vomit and faeces were emitting gamma radiation. Transferred to a private room, Kelley was placed in an oxygen tent and his blood was drawn so that the dose to which he had been exposed could be estimated by measuring his levels of activated sodium-24. The nuclei of sodium and other light metals in the body, when bombarded with neutrons, can acquire extra neutrons. In the case of a sodium atom, this increases its atomic mass from 23 to 24, and this isotope of sodium is unstable and thus radioactive (decaying to magnesium by emitting beta and gamma radiation), and hence is said to have been "activated". The level of activated sodium can thus indicate the intensity of neutron flux to which a person has been exposed.

It was estimated that Kelley had received 900 rads from fast neutrons and up to 4,000 rads from gamma rays, to give a massive total whole-body dose of 4,900 rad (a dose of 1,000 rad is generally considered fatal), equivalent to about 12,000 rem (remember that rads measure exposure while rem measure the biological impact of that exposure). The other two workers who had been in the tank room with Kelley received relatively minor doses of 134 and 53 rem and suffered no lasting ill effects. Kelley, however, had only a few hours to live. Six hours after the accident, it was almost impossible to detect white blood cells in his circulation, and when a bone marrow biopsy was performed 18 hours after this, it revealed a watery, bloody ooze of dead cells and fat; the intense radiation had completely destroyed his vulnerable bone marrow. The pain in Kelley's abdomen grew unbearably severe and he began to sweat uncontrollably and turned ashen grey. Despite sedation he was extremely restless, continually dislodging intravenous drips. His pulse became irregular, and 35 hours after the accident, Kelley died.

Back at the DP West facility, Kelley's accident prompted the planned replacement of the mixing tanks to be brought forward. The squat tank that had killed Kelley was ripped out, and along with the others was replaced with the new, tall column design. Six weeks after Kelley's death, the facility reopened and plutonium processing restarted.

For Kelley's battered body, death was just the start of a gruesome odyssey. Kelley was an object of extreme interest for the health physicists, not just because of the colossal dose to which he had been exposed, but because, as a long-serving employee of Los Alamos, with an accumulated six years of service in plutonium processing, he had regularly been exposed to small doses of plutonium. In the words of a Los Alamos review, "his tragic death became an opportunity to determine certain factors crucial to the protection of workers". In other words, his body represented a rare natural experiment in the physiology of plutonium exposure, which scientists now had the opportunity to dissect. In particular they wanted to measure the total amount of plutonium in his body, and compare it with assays that had been performed on Kelley's urine samples at various stages in his career, and they wanted to see where in his body the plutonium had ended up. It turned out that the total amount of plutonium estimated to be in his corpse – 18 nanocuries – matched up almost perfectly with the 19 nanocuries estimated as his lifetime exposure, and that 50 per cent of the metal had ended up in his liver, 36 per cent in his skeleton, 10 per cent in his lungs and 3 per cent in his respiratory lymph nodes. The health physicists were even able to track how long it had taken plutonium to migrate around Kelley's body, because the isotopic makeup of plutonium processed at Los Alamos had changed slightly over the 11 years of his service.

This detailed work-up of Kelley's tissues marked the beginning of the Los Alamos Human Tissue Analysis Program (HTAP), and involved the removal of around 4 kilograms (9 pounds) of tissue. The only problem, according to a lawsuit filed by Kelley's family in 1996, was that no one bothered to ask them. Although Kelley's widow gave permission for an autopsy, she did not suspect, when, after the procedure was complete, she received the sealed casket for burial, that the body within was missing significant portions. "[Los Alamos] took most, part or all of internal organs, muscles, tissue and bone and shipped it to labs around the country," alleged a lawyer acting for Kelley's widow and daughter. The response of the laboratory was to dismiss the accusations of bodysnatching for a secret project. The HTAP, they claimed, was entirely open, involving samples collected from 1,520 people, all of whom had given consent. Furthermore, they pointed out, the autopsy consent form specifically "allows pathologists to remove for diagnostic, scientific or therapeutic purposes tissues as judged to be proper". It is not clear what became of the lawsuit, but America's nuclear programme has seen off decades of such legal attacks, so the Kelleys are unlikely to have progressed far.

Kelley's fate was not, unfortunately, unique. In 1964, a similar accident befell Robert Peabody, an experienced chemical engineer working at a uranium reprocessing plant in Wood River Junction, Rhode Island. Just as Kelley had done, Peabody received a colossal blast of radiation (as much as 10,000 rad) from an excursion when a non-critical solution achieved a critical configuration after being added to the wrong-shaped container. And in a strange echo of the congruity between the Windscale fire and the Kyshtym Disaster at the Mayak plant, Kelley's accident also had a near-contemporaneous parallel at the Mayak facility. Almost exactly a year before Kelley was killed by an improper vessel configuration, three Soviet workers also found, to their cost, how easily an excursion could be triggered by disregarding the geometry of a fissile fluid.

The incident occurred on 2 January 1958, on the first day back at work after the holidays, as an experienced team performed experiments designed precisely to prevent the accident they were about to cause. The Mayak plant had already experienced a number of incidents with fissile fluids, including at least two prior criticality accidents, and there were concerns that the vessels in widespread use in the facility were of "unfavourable geometry", in the words of the *Los Alamos Review of Criticality Accidents*, the primary source for information on this accident. Accordingly plant managers had contrived a set-up in which a large steel tub, 0.75 metres (2 feet 6 inches) across, was bolted to a stand and connected to various pipes, tubes, detectors, monitors and other equipment, including neutron sources. Solutions of highly enriched uranyl nitrate, of varying concentration, could be fed into the vessel and prompted to varying levels of reactivity and criticality, while the operators worked at a control panel hidden safely behind a 0.5 metre (1 foot 6 inch)-thick, water-filled concrete slab tank. The experimental protocols strictly prescribed that, once an experiment was complete, the tank should be carefully drained via a discharge line into 6 litre (13-US-pint) bottles of favourable geometry (i.e. tall and thin). This would help to prevent too much fissile material being collected in a single volume, at least while workers might be handling it.

On this occasion the experimenters had completed a test run and begun the laborious and tedious task of draining the contents of the test tank into a series of the bottles. Given that the tank could hold up to 400 litres (106 US gallons), this might involve filling scores of bottles. After filling an unrecorded number of the bottles, the operators decided that the solution remaining in the tank must now be well below the criticality threshold, and that extracting 6 litres (13 US pints) at a time was simply a waste of time. Accordingly they decided to disregard the protocols and just empty out the tank by hand. Three of them disconnected the equipment hooked up to the tank and unscrewed it from its stand. Together they hauled it to the edge of the stand, tipping it slightly as they prepared to lower it to the ground.

At that moment there was a blue flash and the contents of the tank were violently ejected, splattering on to the ceiling 5 metres (16 feet) overhead. The three operators dropped the tank, and together with a fourth operator who had been standing 2.5 metres (8 feet) away, went straight to the change room, showered and were taken to hospital. The three who had been holding the tank died five to six days later; it was estimated that they had received doses of 6,000 rad and perhaps as many as 8,000 rad. The fourth worker, who received an estimated 600 rad, survived but suffered acute radiation sickness and long-lasting health problems thereafter, including developing cataracts and losing the sight in both eyes some years later.

Poor monitoring and record-keeping, together with the tight secrecy that surrounded Soviet nuclear operations, mean that many of the details of the accident remain obscure. In an intriguing update of the "phantoms" (blood-filled dummies) used to model the 1946 Demon Core incident (see page 26), scientists from the US nuclear weapons design agency Sandia National Laboratories created virtual phantoms while computer modelling the Mayak 1958 accident in 2015, and used them to help work out some of the details of the accident. One suggestion, for instance, was that the bodies of the three operatives themselves had acted as reflectors, increasing the reactivity of the fissile system as they crowded round it, in similar fashion to Otto Frisch's brush with criticality back in 1944 (see page 10).

The Sandia researchers specifically examined this question. Using sophisticated computer modelling of the dynamics of the fluid in the vessel, they were able to show "that the experimenters had very little effect on the criticality of the system". They concluded that the reactivity added by the reflection of neutrons was "insignificant compared to the primary reactivity addition due to the change in solution geometry in the vessel". In other words, tipping the tank was their fatal error. The very slight change in the geometry of the highly concentrated uranyl nitrate had pushed it over the threshold of criticality from sub to prompt critical.

Why didn't the experimenters realize the danger? How can we account for this lapse of judgement by operatives who were not green new recruits or unskilled labourers, but experienced experts? The answers probably lie in the same toxic stew of stresses and pressures that made Mayak such a dangerous and dirty place to work. Faina Kuznetsov, who worked at Mayak from its opening until 1956, recalled that

> [one] reason why we had a huge number of accidents was because we worked in a terrible rush, and the problem was compounded by our being required to work in strict secrecy . . . all work was under the control of KGB agents. Any delay or mistake was immediately punished. Very often fear pushed people into doing things that caused accidents.

The 1958 incident operatives supposedly had a detailed checklist of operations to follow, but according to Kuznetsov there was a prohibition against any form of written protocols at Mayak: "All the details and different stages in the production processes had to be kept in memory to avoid the possible loss of strictly classified information. Workers were always in a state of stress lest they forget to perform some important step in the process."

Above all, operations at Mayak were dominated by the pressure to get results and meet work quotas. "Anyone who worked at Mayak," Kuznetsov recalled,

> regardless of his position or accomplishments, could be punished at any time. When you worked 'under war-time conditions' (always the case at Mayak), you were under greater pressure than if you were in a labour camp. If you followed the safety rules, you couldn't meet your work quota. On the other hand, if you violated the safety rules you risked losing not only your bonus money, but your life.

Another Mayak veteran, Victor Slavkov, pointed out that, officially, there was no such thing as a nuclear accident – a term that did not appear in the official Soviet lexicon until after the Chernobyl disaster. "We described nuclear accidents as 'leakages,' 'spills,' 'crumbles,' 'disperses,' 'hotbeds,' and 'slaps,'" Slavkov recalled. "And it was never explained or documented that when we used those terms we were talking about the loss of control of radioactive materials."

Accidents such as Kelley's mixer mishap or the overtaxed tank tippers of Mayak revealed with brutal clarity the cost of disregarding and delaying safety measures. Yet there is evidence that even today similar mistakes may be being made in the very same facilities. At Los Alamos, for instance, the plutonium-handling facility, PF-4, is at the centre of a storm over chronic safety problems. An alarm in 2011, triggered by operatives lining up eight rods of plutonium next to one another so they could take a photo, unaware that they were a hair's breadth from a critical excursion, set in motion a chain of events that eventually saw PF-4 shut down for more than four years from 2013. Even today safety concerns persist. The most recent annual report to Congress by the Defense Nuclear Facilities Safety Board, in March 2020, concluded that, with regards to PF-4, "significant portions of the Department of Energy's strategy to upgrade the safety controls have been delayed and the upgrades remain incomplete". Hopefully there will never be another tragedy like Cecil Kelley's but such shortcomings make this possibility disturbingly more likely.

Chapter 6

A SLIP OF THE HAND? IDAHO FALLS, 1961

On the night of 3 January 1961, three young soldiers were killed when a small military research reactor located near Idaho Falls, Idaho, briefly went supercritical. Within just a few milliseconds the incident generated a massive burst of radiation and enough explosive power to lift the entire reactor vessel over 2 metres (almost 9 feet) into the air, blasting the soldiers to death and leaving one impaled on the ceiling. The cause of this devastating accident? A simple slip of the hand as one of the unfortunate grunts man-handled a stuck control rod, resulting in it lifting a couple of feet clear of its intended range – just enough space to unleash the terrifying destructive power of criticality. Behind this immediate cause, however, lay a shabby history of poor design, shoddy maintenance, penny-pinching management and a shocking lack of oversight, a catalogue of failings at risk of being concealed by a tawdry tale of marital stress and crimes of passion.

The atomic bomb had been the main focus of America's nuclear research programme during and immediately after the war, and the US had been relatively slow to develop nuclear power in comparison with the Soviets and the UK. But the military applications of nuclear power were clear: nuclear reactors could provide self-contained, long-lasting power and heat generation using fuel with a higher energy density than any other substance. This logic had driven the successful development of nuclear reactors for the American submarine fleet, and the US Army was pursuing its own research programme, including the development of small reactors that might provide power and heat for Arctic installations.

The prototype for this role was Stationary Low-Power Reactor 1 (SL-1), an experimental reactor installed at the National Reactor Testing Station (NRTS, later the Idaho National Laboratory) in the military settlement of Arco, in remote eastern Idaho about 65 kilometres (40 miles) west of Idaho Falls. SL-1 was a 3-megawatt thermal boiling water reactor, in which heat generated by fission made steam, to drive a turbine-generator to produce electricity, and a condenser to provide heating. The reactor itself looked a bit like a giant steel test tube: a round-bottomed steel tube roughly 4.5 metres (15 feet) tall, capped with a shield, and an assembly housing the drive mechanism for the control rods (see below). Inside this tube was the reactor core, containing 14 kilograms (14 kilos rounds up to 31 pounds) of 93-per cent-enriched U-235 in 40 aluminium fuel assemblies. The tube was half filled with water, with the core sitting in the water and heating it into steam, like the element in a giant, radioactive electric kettle.

The design of the reactor seems crude by modern standards, but this was the point of it. The US military wanted a reactor design that was simple, cheap, robust and, most importantly, did not rely for its operation and management upon a cadre of highly educated, specialized and extremely expensive nuclear engineers and scientists. SL-1 was an experimental design, and one of the factors that was being tested was the resilience of nuclear power in the context of the US military.

The reactor first went critical on 1 August 1958, and successfully supplied most of the energy demands of Arco for the next two and a half years. But this apparent track record of success concealed a disturbing catalogue of red flags in the design, control, maintenance and operation of the system, which made an accident likely if not inevitable.

Although the government agency then known as the Atomic Energy Commission (AEC) had overall responsibility for SL-1, the operation of the reactor was outsourced to a contractor. In February 1959 this role had been taken over by Combustion Engineering (CE), who used young military personnel to perform the actual hands-on work. These under-prepared young men worked in three 8-hour shifts but only the daytime shift was overseen by supervisory staff; the AEC had refused CE's request to fund additional supervisory staff. The three young men working the evening shift shortly after New Year, 1961, thus had no one monitoring their work, keeping records or watching out for warning signs. Instead they had a demanding list of tasks they needed to get through as the reactor was brought back on line following its usual holiday shutdown.

One of the many compromises that had been made in the design of the reactor was to strip right back to basics the monitoring mechanisms. For instance, there were no mechanisms for active monitoring of the flux profile of the reactor – the way that neutron production varied across the geometry of the reactor. Instead there were "flux wires", aluminium wires with pellets of cobalt, set at intervals around the reactor. In order to "read" the information these wires gathered, the wires had to be physically removed and processed, and this required the complete shutdown and partial dismantling of the reactor. This was the procedure that had been carried out before Christmas, and the task of the three men on the graveyard shift was to help complete the laborious task of putting the reactor back into working order. Specifically, they needed to reconnect the control rods to their drive mechanism.

The control rods were cross-shaped, cadmium-enriched blades that could be raised and lowered. When fully inserted, the cadmium rods soaked up enough neutrons to prevent the radioactive decay of the U-235 fuel from sustaining fission chain reactions (see page 363). Raising them a few inches would allow sufficient neutrons through to trigger enough fission events to generate heat, boil water into steam and drive the turbines. There were five control rods, with four around the outside and one in the centre; a poorly designed set-up, since it meant that the centre rod disproportionately affected the reactivity of the fuel core.

One of the main factors in governing the vulnerability of a reactor to a meltdown is the "reactivity worth" of the various elements – particularly the control rods. An element with a high reactivity worth can have a disproportionately large influence on the behaviour of a reactor, so that if something goes wrong with that element, the risk of an accident is disproportionately elevated. This was exactly the situation in the SL-1 reactor, where the poor design of the control rod geometry meant that the central control rod had a disproportionately high reactivity worth. Simple geometry shows that what might be called the 'zone of influence' of a rod through the centre of a mass, is greater in size than that of rods on the outside, since the central rod can influence the core on all sides, while the outer ones can only influence the core to the inside of them.

Normally the raising and lowering of the control rods was effected by the drive mechanism, but during shutdown the tops of the rods had been disconnected from the drive, and in order to bring the reactor back on line they had to be reattached. An upgrade scheduled for the spring of 1961 was intended to automate all the steps, but in January this task still had to be carried out by hand, which was why on that fateful night two men, 26-year-old Richard Legg and 22-year-old John Byrnes, had climbed on top of the reactor to wrestle with one of the world's most dangerous machines. Disastrously for Legg, Byrnes and their co-worker, 27-year-old Richard McKinley, the control rods they needed to lift were almost certainly stiff – possibly even stuck – due to the poor state of the reactor.

The operation of the control rods was just one of several fundamental design flaws in the reactor; another was that many of its components would degrade and warp under the stress of operation. One example was the system of boron strips welded to the fuel assemblies, which served as an additional reactor control element, with the boron acting as a "neutron poison", a substance that absorbs neutrons and thus limits the ability of the reactor fuel to sustain a fission chain reaction. Such poisons would later play a crucial role in the Chernobyl disaster (see page 251).

Exposure to the energies and particles generated by a nuclear core, including constant neutron bombardment and high temperatures, can be extremely damaging to materials, and the boron strips in the SL-1 reactor were no exception. They had bowed as they deteriorated, which made it difficult to remove the fuel assemblies for inspection. As a result, according to Tami Thatcher, a former nuclear safety analyst at the Idaho National Laboratory and nuclear safety consultant who conducted a detailed review of the incident, inspection of the fuel assemblies stopped. By the time of the disaster, the boron strips had deteriorated to such an extent that, apart from no longer adequately performing the reaction damping function for which they were intended, they were flaking and buckling extensively, and this was probably one of a series of contributing factors to the fatal "stickiness" of the control rods.

As a result of inadequate monitoring and poor record-keeping, there was no way to know if the control rods were even in the correct position. In addition, as with the boron strips, the shrouds around the control rods suffered distortion caused by long exposure to radiation. Defects exacerbated one another: flakes from the deteriorating boron strips were shed as debris that could impede the movement of the rods. Design flaws in the shrouds for the rods meant that there were at least two ways in which such debris could get inside. When the reactor tank was open for maintenance, it was possible for material to be dislodged, fall into the tank and get into the shrouds. What is more, the shrouds had "weep holes" about 5 centimetres (2 inches) in diameter, through which debris could enter, while the followers (the bits on the end of the control rods) were known to have weld ridges and discontinuities that could catch on the edges of distorted weep holes or debris.

All of the problems with deterioration of the materials in the reactor were further exacerbated by an aggressive programme of high power testing that had been run in late 1960 (in order to test a new steam condenser), prior to the holiday shutdown. Running a core at high power generates more neutrons, more radiation of other types and more heat, all of which can damage and corrode materials. This was yet another example of how poor management had primed the reactor for disaster, and warning signs were evident, because the control rods had got stuck before – not once, but dozens of times. The centre control rod alone had got stuck on seven previous occasions, while the frequency of such occurrences was increasing. Up to 18 November 1960, control rod problems occurred only 2.5 per cent of the time; between 18 November and 23 December, this leapt to 13 per cent. When the reactor was shut down on 23 December, only two of the five rods dropped cleanly to the bottom of the core, as they were supposed to. After the accident, investigators would find "scouring" marks on several of the control rods, indicating the degree to which they had scraped and stuck in their shrouds.

Further evidence of the fatal synergy of poor design and management includes the changes made to the reactivity shutdown margin – effectively the margin for error in terms of keeping under control the nuclear reaction and its attendant neutron-production. Radiation-linked degradation of elements was already known to be adversely affecting the shutdown margin. Accordingly it was decided to boost the damping precautions attached to the fuel elements, by inserting extra "poisons" in the form of cadmium strips (known as "shims") to be welded to the sides of the reactor. The calculations performed to assess the effectiveness of this step – and ensure the safety of the reactor – were based on the assumption that shims would be added on four sides of the reactor, but in the event they were added on only two sides, and even some of those strips were installed slightly too low. These factors made the reactivity of and subsequent power load in the reactor asymmetrical, with serious consequences for its vulnerability to "flash" heating.

Another in the catalogue of errors that contributed to the volatility of the reactor was that one of the central – and disproportionately reactively "valuable" – fuel assemblies was replaced at some point with a relatively fresh assembly. As fuel assemblies accumulate time in the reactor they burn through their fuel and their potential power output decreases. Replacing an older assembly with a newer one thus disrupts the symmetry of elements in the core, while also increasing the potential to generate a high number of neutrons.

Neutron flux is the key determinant of the reactivity – and potential threat – of a reactor core. More neutrons mean more fission reactions, generating in turn even more neutrons and greater energy release. But just as important as the total neutron flux, in determining the course and result of an accident, is the rate at which the neutron flux increases. When the neutron flux jumps extremely rapidly, so much heat is generated so quickly that there is no time for it to radiate away from the fuel (into the coolant water). Instead the temperature of the fuel instantly rockets and the fuel melts: the infamous meltdown, which can have catastrophic consequences. In the SL-1 reactor, as the result of all the missteps, design flaws and problems outlined above, the "reactor period" (the time interval required for the neutron flux or density to more or less double) had fallen to just 4 milliseconds in the event of a sudden increase in reactivity. So when a mistake was made, as Legg and Byrnes were about to discover while grappling with the central control rod, the consequences would be almost instantaneous.

How could management have let the SL-1 reactor get into such a state? The SL-1 reactor was specifically designed to test the proposition that a small nuclear power unit could be operated and managed by non-specialist, relatively junior military personnel. Nuclear historian James Mahaffey suggests that, when a mishap or problem occurred that would normally flag the need to shut down and decommission a reactor, the authorities saw it "as an interesting perturbation thrown into the exercise". He describes their attitude as, "Let's see how it plays out", and suggests that they wanted to simulate the real-world context in which such a device would be deployed. SL-1-style reactors were intended to be used at remote bases, where experts and resources would not be close at hand. Could ordinary grunts manage a basic reactor on their own? Would the relatively simple design of the SL-1 prove robust enough to cope with wear and tear? Both questions would be definitively answered on 3 January.

On the night of the accident, the three men on duty were the more experienced Richard Legg and John Byrnes, and a novice, newly transferred from the Air Force, Richard McKinley. Legg and Byrnes were known to be an unhappy pair. Legg was touchy and sensitive to slights and liked to play practical jokes on colleagues. Byrnes was having issues with money and with his marriage, which was breaking down. He also had concerns about the reactor, which had already shown signs of erratic behaviour. On 23 November the previous year, an automatic scram (rapid shutdown of the reactor by fully inserting all the control rods at once) had been triggered when it experienced a sudden power surge, moving Byrnes to comment to his wife that he was worried the reactor might blow up.

As the new boy, McKinley was briefed simply to observe the other two work through their list of operations, and to monitor radiation levels with a portable ionization detector nicknamed the "cutie-pie", thanks to its diminutive size. Legg and Byrnes had printed instructions to follow. The central control blade had been clamped a few centimetres clear of its lowest position, so that a metal rod could be screwed into the top of it, allowing manual raising of the blade. Legg's job was to unscrew the clamp, whereupon Byrne was supposed to raise the control blade by just 2.5 centimetres (1 inch), taking the load off the clamp so that Legg could remove it, after which Byrnes was to lower the blade so that its bottom rested on the floor of the reactor.

For some reason, this programme of events was not properly followed, and one of the men – exactly who is still disputed – jerked the control blade too far: according to the official account, a full 58 centimetres (23 inches). With the disproportionately reactively valuable cadmium control blade lifted clear, the core achieved what nuclear physicists call prompt criticality (see page 366–367). In the space of just a few milliseconds, an uncontrolled fission chain reaction generated 2 quadrillion neutrons, and a 15 megajoule burst of energy, roughly equivalent to the energy of a 15-tonne truck smashing into you at 160 kilometres per hour (100 miles per hour). The superheated core flashed the surrounding water into steam, unleashing terrible forces as the plug of water above the core was driven upwards towards the top of the reactor vessel. The water hit the shield on the top of the reactor like a colossal hammer, impacting with a peak pressure of about 10,000 psi. The impact lifted the entire 13-tonne steel reactor vessel 2.7 metres (9 feet) into the air, shearing off the pipes and tubes at the bottom. Plugs in the shield shot out like bullets, and pieces of control rod were buried deep in the ceiling of the reactor room.

Byrnes and Legg were killed instantly by the shockwave, and they were already dead when Byrnes was blasted off the top of the reactor and Legg was pinned to the ceiling by a piece of debris. McKinley, who was standing off to one side, was slammed against a concrete block and suffered a massive head wound. He was still alive when a fire crew attended the scene but died two hours later (although some sources identify this victim as Byrnes). If the shockwave had not killed them, all three men would have died from massive radiation damage, both from the critical flash, and from the radioactive material driven deep into their tissues. The autopsies on their bodies had to be performed from behind lead shields, with instruments on the ends of 3 metre/10 foot-long poles, while their remains were later buried in lead-lined caskets. Not all of the remains, however, because some body parts were buried at the NRTS, along with the damaged reactor itself, which was later interred in a massive trench as part of a clean-up operation that took 13 months and cost $2.5 million (adjusted for inflation, roughly $21.5 million in today's money).

Part of the problem was that radioactive debris was scattered all around the inside of the reactor building, and locating every particle of it – including tiny fragments deeply embedded in the ceiling – would require some ingenuity on the part of the investigators. One clever trick they used was a gamma-ray pinhole camera. A pinhole camera is the simplest possible form of camera: a box with a pin-sized hole in the side that admits just enough light to project a view of the exterior on the side of the box opposite the hole, where a sheet of film can be placed to record the view. Exposure of the film is controlled simply by uncovering and then re-covering the pinhole. A gamma-ray pinhole camera is exactly the same, except it is used in the dark, and the box and hole cover are impassable to gamma rays. When the pinhole is uncovered, gamma rays can enter the box and expose the film; the brightest gamma ray sources – such as radioactive debris buried in the ceiling – will register as the most exposed points on the film. Once the film was developed the resulting picture could be compared with the ceiling of the reactor building to locate the hidden debris.

Incredibly the steel shell of the building had withstood the impact of the explosion and contained the damage, so that when the fire crew, summoned by alarms, arrived, they could see no sign that anything was amiss. Piecing together the course of events would take months, and the official account would eventually lay the blame squarely at the door of "operator error" – in other words, one of the men had acted in such a way that an otherwise inconceivable accident had occurred, in a reactor the design of which was otherwise sound.

At the same time, reporting of the event took a darker turn, with suggestions that the incident was, perhaps, not an accident after all. Byrnes's marital woes were unearthed, and it was insinuated that he had overextended the control rod on purpose, an interpretation for which the authorities had carefully left space with their statement that "we may never know why the rod was lifted too high". Similarly, the AEC were deliberately coy in their take on the motive, explaining in a film about the incident:

> Direct cause of the accident clearly appears to have been manual withdrawal of the central control rod blade by one or more of the crew members — considerably beyond the limits specified in maintenance procedure. However, there was insufficient evidence to establish the actual reason for such abnormal withdrawal.

The suggestion was that Byrnes perhaps believed his wife was having an affair, and that he was suicidally distraught over being caught in a love triangle. The elaboration of this lurid narrative was assisted by the initial confusion over the identities of the three victims, and convenient obfuscation about which of them had committed the fatal act of pulling on the control rod. Initially it was believed that the man impaled to the ceiling was the new trainee, Richard McKinley, and that the incident must have been the result of a mistake by the inexperienced novice. When it transpired that it was actually Richard Legg who was impaled to the ceiling, a new narrative emerged, based on the unfounded implication that Legg was somehow involved in John Byrnes' marital problems, and suggesting that Byrnes had perhaps arranged this grisly fate by deliberately overextending the control rod, in a desperate act of murder-suicide. Even if this were somehow possible, the evidence seemed to indicate that it was not Byrnes who had handled the rod, as his hands did not show the signs of damage and contamination consistent with this scenario. Even today, however, many reputable accounts of the incident say it was Byrnes who pulled the rod.

Did the rumour-mongers stop to think about the implications of their innuendo? Exposing yourself to a high neutron flux would have to rank among the most unpleasant and painful ways to commit suicide, although it is perhaps unlikely that Byrnes or the other men would have been fully informed about the horrors of such exposure. It is also unlikely that they would have been alert to the dangers of such an incident. The boiling water reactor design of the SL-1 had been touted by the AEC as "inherently safe". It was argued that in such a reactor, "the accidental addition of any amount of excess reactivity . . . can be removed by the formation of steam before the power rises to a dangerous level". The possibility this belief might be mistaken, and an explosion might result, was deemed so unlikely that military investigators initially suspected the incident had been caused by a bomb, planted as an act of terrorism or sabotage.

Nuclear historian James Mahaffey, who maintains that Byrnes was the one who pulled out the control rod, suggests that he did so to show off or to prank the new guy, McKinley, an Air Force man. The Air Force was known to be running exciting nuclear experiments down the road from Arco, while the whole point of SL-1 was to be prosaic and routine, so it is conceivable that Byrnes had a chip on his shoulder. Mahaffey speculates that he wanted to give McKinley a fright – "a thrilling blip" – by generating a momentary burst of neutrons that would set his cutie-pie detector buzzing. The tragedy, Mahaffey suggests, was a prank that went wrong.

Given everything that is now known about the condition of the central control rod, the potential uncertainty over its starting position, and the enhanced vulnerability of the reactor to a supercritical incident, it seems possible that the actual distance the rod was withdrawn may have been significantly less than the extreme suggested by the official account. And it seems highly likely that such a withdrawal was accidental, not deliberate – the result of an insufficiently trained, unsupervised and overworked non-specialist yanking hard to free a warped control rod snagged on a piece of debris.

Why was such a scenario not reflected in official accounts? It seems possible – perhaps even likely – that the authorities were happy for the rumours to be accepted as fact. Better for the actions of a rogue lone operative to be at fault, than to have to own up to the truth about shoddy maintenance and lack of oversight compounding the risks of an inherently unsafe design. Tami Thatcher concludes:

> The complexity of the accident and the long months of investigation would play a role in the speculation of what caused the accident. But, carefully crafted statements had two objectives (1) divert blame from the AEC and its contractors and (2) do nothing to undermine public faith in the nuclear industry.

Chapter 7

BROKEN ARROWS: NUCLEAR WEAPONS ACCIDENTS, 1958–68

IN 2010, writing in the *Bulletin of Atomic Scientists*, Robert Norris and Hans Kristensen estimated that there were almost 22,400 intact nuclear warheads in the world, of which nearly 8,000 were operational to some degree and 1,880 were at an alert level of readiness (ready to launch at short notice). This is significantly less than the peak number of warheads, reached in 1986, when there were 69,368 warheads in the world. Since 1945, they estimate, more than 128,000 nuclear warheads have been built. Remarkably, not one has detonated by accident, but there have been dozens – probably hundreds – of accidents involving nuclear weapons, including many that caused radioactive contamination and/or resulted in the misplacement, and sometimes permanent loss, of weapons-grade fissile material. In many of these incidents the nuclear bomb *has* exploded, in the sense that the shell of high explosives that implodes the fissile material has indeed detonated.

Fear of premature, accidental detonation, or of some other catastrophic mishap, dates back to the very first atomic bomb: the Gadget set off in the Trinity test. The scientists behind the bomb could not be entirely certain that its successful detonation would not ignite the Earth's atmosphere, and indeed they made wagers with one another about this very outcome as a form of mordant humour. More seriously, there was great concern about the possibility of premature activation of the complex electronic detonation system. The Gadget was an implosion-style bomb; its operation relied upon the perfectly synchronized detonation of a geodesic sphere of high explosives. To accomplish this feat, Manhattan Project scientists had invented a new kind of detonator, in which a thin silver wire was vaporized by the passage of a high-voltage electric current. This current was supplied by a contraption designed by Donald Hornig, at the age of 25 one of the youngest scientists at Los Alamos. Known as the X-unit, it comprised a bank of capacitors that could discharge current to multiple detonators simultaneously. The Gadget required a pair of these wired up to the explosives with a complex spaghetti of cabling that gave the whole assembly a sci-fi aesthetic worthy of any mad scientist.

What was particularly concerning was that, a week before the Trinity test, on 9 July, an X-unit had been triggered into a premature discharge by the build-up of static electricity during a lightning storm. "That made everyone nervous," recalled Hornig many years later. Could the same thing happen when it was attached to the Gadget? What would happen if the device were hit by a lightning strike? No one knew for certain, but the prospect was alarming, even more so when the weather during the Trinity test window deteriorated, with a succession of storms passing through. On 13 July 1945, the plutonium core for the Gadget, which had been assembled by Louis Slotin and others at the nearby McDonald ranch house (see page 33), was delivered to a large tent at the base of a 33 metre/100 foot-high metal tower in the desert. Here it was to be inserted into the centre of the shell of high explosives, themselves encapsulated by an aluminium casing, around which snaked the cabling from the X-units. At about 16:00 a thunderstorm threatened to break overhead, and the bomb assemblers beat a hasty retreat to the ranch house for an anxious half-hour wait as the storm passed.

Anxieties mounted still further the next morning, as the Gadget was hoisted into place atop the tower. A great heap of army mattresses was piled underneath it, in case the cable snapped. Once in place, the bomb waited for a window in the weather that would allow it to be set off without risk from lightning strike, or from high winds carrying radioactive fallout far beyond the test region. Robert Oppenheimer, the man in charge at Los Alamos, was paranoid about the risk of sabotage; he wanted, Hornig says, "someone who understood the details" to "baby-sit the live bomb . . . until it was locked up for firing, and as the youngest, I had priority."

At 21:00, in the rain, Hornig climbed the tower to a metal shed at the top, in which sat the Gadget. It was fully armed and ready to go. Hornig knew better than anyone how the X-units might respond to atmospheric static electricity, and now a violent electrical storm blew up close by. "It was a deeply philosophical experience," Hornig remembered sixty years later. At midnight he was ordered to climb down, and a few hours later the weather cleared and the test went ahead without being derailed by either accidental detonation or sabotage. The first nuclear weapon had not gone wrong. The problem for the new nuclear weapons industry was the absolute imperative of maintaining this 100 per cent success rate.

This imperative would be severely tested in the theatre of war, as the new A-bombs made the transition from research project to deployment. The bombs – Little Boy and Fat Man – were to be carried to their targets by US Army Air Force (USAAF) B-29 bombers taking off from the airbase on Tinian Island in the Pacific, over 1,600 kilometres (1,000 miles) from Japan. The arming and fusing of each bomb was a three-stage process. Arming wires were secured to the bomber and thus stripped out when the bomb was released, at an altitude of 9,150 metres (30,000 feet); this triggered spring-wound mechanical clocks to count down for 15 seconds, at which point they would close a switch and connect the firing circuit to the bomb's battery. Meanwhile another set of switches were controlled by barometric sensors that responded to changing air pressure such that, at an altitude of 2,133 metres (7,000 feet), they would activate ground-sensing radar units. These in turn would close a switch when they sensed that the distance to the ground was 564 metres (1,850 feet), sending the signal to fire, which either meant, in the case of Little Boy, detonating cordite bags to fire a uranium projectile at another piece of uranium, or, in the case of Fat Man, triggering the X-units.

In other words, if everything worked, the bombs should not go off until they were about 500 metres (1640 feet) above a Japanese city. But Oppenheimer was worried enough to write to his USAAF liaison in 1944 to ask, "whether the take-off can be arranged at such a location that the effects of a nuclear explosion would not be disastrous for the base and the squadron." Each bomb type had its particular risks attached. It was feared that the explosive lenses of Fat Man might all too easily be accidentally set off by a bullet or a fire or some other unpredictable trigger. The cordite explosives of Little Boy might be similarly vulnerable to accidental detonation, while the experiments at Los Alamos on submersion of enriched uranium in water (see page 11) had demonstrated an additional risk factor: if the bomb had to be dumped into the ocean, or if the plane carrying it had to ditch, seawater could trigger a supercritical excursion even if the explosives did not go off. A Target Committee set up by the White House, the remit of which included such considerations, was forced to conclude that, for Little Boy, there was "no suitable jettisoning ground". The best advice they could offer the bomber crew was to try manually to disarm the bomb in mid-air and make sure they crashed on land.

Such considerations prompted an extraordinary decision by Captain William Parsons, the bomb commander and weaponeer for Little Boy, the first A-bomb mission. Parsons was the chief ordnance officer of the Manhattan Project and had overseen development of the gun-type bomb; he was all too aware of the risk of accidental detonation. Accordingly he decided, without telling General Leslie R. Groves, head of the Manhattan Project, that he would complete assembly of the bomb only after the B-29 *Enola Gay* had taken off and was a safe distance from Tinian. This necessitated himself and weaponeer Morris Jeppson climbing into the bomb bay once the plane had flown about 95 kilometres (60 miles) from the airbase and climbed to 1,500 metres (5,000 feet), and inserting four small bags of cordite into the breech of the gun barrel, as the bomb hung from a hook and the aircraft was buffeted by turbulence. About four and a half hours later, as they neared their target, Jeppson returned to the bomb bay and replaced the green safe-ing plugs, which blocked the circuits between the fuses and the cordite charges, with red arming plugs. At 8:15 on 6 August 1945, Little Boy fell from the bomb bay, dropping away from the arming wires, setting in motion the timer for the arming switch. After about 40 seconds the barometric switches turned on the radar units, and about four seconds later these in turn activated the firing signal. Little Boy exploded over Hiroshima with an explosive force of around 15 megatons, destroying two-thirds of the buildings in the city and killing 80,000 people.

The preparation of Fat Man, a few days later, also involved unplanned risks. The night before the bomb was to be loaded onto a B-29, technician Bernard O'Keefe noticed that the cable which was supposed to connect the radar units to the X-units had been installed backwards. The two ends of the cables which were supposed to connect together were both female. Correcting the problem according to the rules would take days. Disregarding a host of safety regulations, O'Keefe brought a soldering iron into the bomb storage room and rewired the cable, wielding a hot iron in close proximity to over 2 tonnes of high explosive surrounding a plutonium core. His gambit paid off and Fat Man was duly loaded aboard the B-29 *Bockscar*. Even then there was a fright when, four hours into the flight, a box in the cockpit flashed red lights indicating that the bomb was armed and ready to detonate. This proved to be a false alarm caused by faulty settings. At 11:02 on 9 August, Fat Man exploded over Nagasaki with a force of 21 kilotons, killing 40,000 people.

Amidst the conflicting emotions felt by the weapons designers in the aftermath of the first atomic bombings, there was considerable relief that the bombs had detonated successfully. The first imperative of nuclear weapon design, then as now, was that the weapon should always go off when intended. The opposing imperative was safety – the necessity that a bomb should never go off unintentionally.

This tension between always/never would come to define the nuclear weapons programme of the US and all other nuclear powers.

One of the simplest ways to ensure a "never" level of safety was to keep the different parts of the bomb separate until the last moment: the "shell" of the bomb and the pit. The shell comprised the explosive lenses and the rest of the bomb, such as arming and trigger devices, while the pit comprised the plutonium core, its coating and a layer of tamper. Early atomic bombs were engineered so that a panel in the side could be opened, and the pit slotted into place manually, although this was a demanding task for weaponeers in the cramped and hostile environment of a bomb bay at 9,150 metres (30,000 feet). Later the process was automated, which reduced the risk of accidents and improper insertion, helping both the "never" and the "always". Although even without the core, the rest of the bomb often included radioactive elements, such as uranium tamper, in addition to hundreds of kilograms of high explosive, with pits not inserted until shortly before deployment, bombs could be transported and stored without any risk of nuclear detonation. Separate cores could thus mean the difference between a Broken Arrow and a Nucflash.

Native American-style terminology for military nuclear incidents was in use by at least the early 1960s; for instance, a 1962 training film prepared for the US Air Force (USAF) by the Defense Atomic Support Agency was titled *Broken Arrow Procedures*. The precise definition of "Broken Arrow" varied, but in general it referred to a serious accident with a nuclear weapon, but one that was unlikely to risk starting a nuclear war. A Nucflash, on the other hand, was an actual detonation of an American nuclear weapon, which in almost any circumstance was likely to run the risk of making one side or the other think that a first strike had been launched, or at least that the other side might think so. A Broken Arrow encompassed any incident, from an accidental nuclear detonation that did not risk triggering a war, to a non-nuclear detonation of a nuclear weapon, to losing a nuclear weapon: anything that caused or risked harm to the public. Other terms included Empty Quiver (theft or loss of a nuclear weapon); Bent Spear (damage to or mishap relating to a nuclear weapon that did not threaten harm to the public or result in any sort of detonation); and Dull Sword (a malfunction rendering a nuclear weapon inoperative).

Separate cores proved to be the saving grace for the US military in several Broken Arrow incidents, most notably the February 1958 Tybee incident and the March 1958 Mars Bluff incident. In the former, a B-47 loaded with a Mk-17 thermonuclear bomb was taking part in an exercise on 5 February, in which it was pretending to be an enemy bomber menacing a plutonium production plant in South Carolina. The B-47 made a dummy bombing run over the plant and turned away. Meanwhile two F-86 Sabres, which had been scrambled to intercept the bomber, were approaching, guided by radar on the ground. Something went wrong with the guidance information and one of the

Sabres collided with one wing of the B-47. The pilot of the Sabre, First Lieutenant Clarence Stewart, ejected as his plane broke up. The crew of the B-47 felt a blow to their right wing and looked out to see one of their engines hanging off at an alarming angle. They managed to stabilize the aircraft and got permission for an emergency landing at Hunter Air Force Base in Savannah, Georgia, but first they needed to lighten the plane. They turned and flew over Tybee Island, and over Wassaw Sound, just off the shore of the island, they dropped their Mk-17 from a height of 2,225 metres (7,300 feet).

Since they had been engaged in a training exercise, the bomb had no pit installed; instead there was a lead dummy capsule. Like most thermonuclear bombs of the era, however, the Mk-17 had a two-stage design, in which a fission primary generated neutrons and X-rays to set off a fusion secondary, and this secondary contained significant amounts of uranium and a small amount of plutonium. So even in the absence of the capsule for the primary, and despite the subsequent impossibility of a nuclear explosion, there was still potential for serious radioactive contamination to result from an accident with the bomb. Fortunately, a recent safety innovation in the formulation of the implosion lens high explosives meant that they would not trigger from impact or shock, and so the bomb landed with a splash not a boom.

The crippled B-47 landed safely, and Lieutenant Stewart drifted down to earth without injury. But somewhere off the coast of Tybee Island there was a ditched H-bomb. A huge search operation ensued, involving the US Navy and the USAF Explosive Ordnance Disposal detachment. Personnel searched beaches, and the Navy brought in almost a dozen craft of various types, and a helicopter. Nothing was found, and the search was eventually abandoned on 16 April. Since only a dummy capsule had been installed, the lost bomb was considered to be a low risk, which could be safely left in the silt at the bottom of Wassaw Sound. This was not, however, the end of the matter. Significant confusion and alarm were caused in 1966 when an Assistant Secretary of Defense stated to a congressional committee that the nuclear capsule had been installed and the lost bomb was a functional thermonuclear weapon.

Over the next four decades the lure of the Broken Arrow would attract many treasure hunters, most notably a retired Air Force colonel who claimed to have located the Mk-17 by towing a Geiger counter behind a motorboat. It turned out that he had simply detected the presence of monazite sand, a naturally occurring radioactive mineral found around Tybee Island. In fact, even an intact bomb with capsule inserted would likely be very hard to detect with a Geiger counter because the radiation emitted would be effectively shielded by the casing and the surrounding seawater. The Mk-17 bomb remains lost somewhere in Wassaw Sound to this day.

Just over a month later, a more serious incident saw a bomb dropped on an American family, in Mars Bluff,

South Carolina. On 11 March 1958, a B-47 bomber took off from the Hunter Air Force Base on a training exercise that involved flying across the Atlantic to England. On board was a Mk-6 nuclear bomb, though without its nuclear capsule/core. For take-off the bomb was unlocked from its rack, as was standard operating procedure, to allow the bomb to be jettisoned in case of emergency during this vulnerable stage of the flight. As they climbed to a safe altitude, the operating procedure was to re-engage the locking pin by pushing a lever in the cockpit, which in turn was attached to a lanyard on the pin. But when the co-pilot pushed the lever, the light on the instrument panel remained red, indicating that the pin had not re-engaged. The pilot, Captain Earl Koehler, asked the navigator/bombardier, Captain Bruce Kulka, to climb into the bomb bay and reinsert the locking pin manually.

The unpressurized bay of a bomber climbing to 4,500 metres (15,000 feet) made for an alarming environment, even more so given that Kulka had to take off his parachute in order to squeeze through the access door. Worse still, Kulka did not know where the pin was, or how to reinsert it. After hunting around fruitlessly for 10 minutes, he reasoned that the pin must be above the bomb. The bomb, however, was the size of a car (about 3.35 metres/11 feet long and 1.5 metres/5 feet wide), while Kulka was a short man; he grabbed hold of a handle in order to haul himself up for a good look. Unfortunately, the handle turned out to be the emergency bomb release, and the bomb fell from its rack on to the bomb bay doors, with Kulka either alongside it or lying on top of it – accounts differ. The Mk-6 weighed nearly 3.5 tonnes (7,600 pounds), and the doors almost immediately gave way, dropping the bomb out of the bottom of the plane. Kulka was perilously close to following it but managed to grab hold of something and pull himself to safety.

At 16:34, the bomb landed on a playhouse in the garden of the Gregg family of Mars Bluff, South Carolina. Walter Gregg and his son were in the tool shed, while Mrs Gregg was in the house and her daughters were playing with their cousin in the garden. The high explosive lenses detonated, digging a crater 21 metres (70 feet) wide and 10.5 metres (35 feet) deep. Incredibly no one was killed, and the only person seriously hurt was the cousin, Ella Davies, who had to go to hospital and received 31 stitches. "When the thing fell," she later recalled, "I remember hearing it, it was the whistle of the bomb coming down. I thought it was an airplane or jet flying over." The house was severely damaged.

The crew in the cockpit of the B-47 realized that the bomb had fallen from the plane only when they felt the shockwave from the detonation. They circled back to take photographs, as required by standard operating procedure, and radioed back to base with a special code indicating a lost H-bomb – an event so rare that the base did not recognize the code. As a result, Captain Koehler had to radio the nearest civilian airport and ask them to call Hunter Air Force Base to let them know that "aircraft 53-1876A had lost a device". The flight crew knew that they were in trouble. On landing they were met by armed guards and temporarily locked up. Poor Kulka became known as the "Nuclear Navigator", and he and the other two crew members were reassigned overseas. In the long term, however, the incident did not destroy their careers; for instance, Captain Koehler eventually retired from the Air Force as a lieutenant colonel, while Captain Kulka made the rank of major. In later years, some of the crew even returned to Mars Bluff to pay their respects to the unfortunate Greggs.

In the wake of the accident, the world's media descended on Mars Bluff and alarming headlines circled the globe. They ranged from the merely factual – "Atom Bomb Without Warhead Drops In Mars Bluff Section" in local paper the *Florence Morning News*, to the sensational – "Are We Safe From Our Own Atomic Bombs?" asked the *New York Times*. The Strategic Air Command (SAC) – the powerful wing of the USAF that controlled most of America's nuclear weapons – countered with propaganda pointing out that the Mars Bluff incident involved a "dead A-bomb" and insisting that it was "the first accident of its kind in history".

This was not true. An almost identical accident had occurred a year earlier near Albuquerque, New Mexico, that went unremarked on by the press. Even worse, a nuclear weapon with its core loaded had been involved in an accident at a US airbase in Sidi Slimane, Morocco, shortly before the Tybee and Mars Bluff incidents. On 31 January

1958, a B-47 carrying a Mk-36 bomb was practising runway manoeuvres when a tyre burst and the plane caught fire. Luckily the high explosives did not detonate, but the bomb and parts of the plane melted into a huge puddle of radioactive slag. In the panic, the airbase was evacuated and carloads of airmen and their families raced into the desert in mortal terror of an imminent nuclear explosion. Unlike the Mars Bluff incident, the Sidi Slimane accident was easily hushed up by SAC, but the fact that the bomb had been armed, with its core/capsule inserted, made this perhaps the worst Broken Arrow of the separate core era.

Advances in bomb technology meant that Broken Arrows of the Tybee and Mars Bluff ilk would be among the last of their kind. In 1955 the weapons designers at Sandia National Laboratories, one of the nuclear weapon design shops that took over from Los Alamos, introduced sealed pit weapons. These are self-contained bombs that do not need predetonation assembly; such weapons are supposedly protected by inherently safe design features. For one thing, the new sealed pits utilized a "boosted-fission" design. This is where the plutonium core at the heart of the bomb is actually a hollow sphere, which, on its own, does not contain enough fissile material to go supercritical even in the event of a perfectly spherical implosion. Supercriticality can be achieved only through "boosting" – injection into the central cavity of a mixture of heavy hydrogen isotopes (deuterium and tritium), at the same time as an electrical neutron initiator fires a blast of neutrons into the mix. The initiator neutrons trigger fission in the imploding plutonium, which in turn triggers fusion in the hydrogen mix, which in turn produces more neutrons to help boost the supercritical fission of the plutonium. All this constitutes the fission primary of a thermonuclear or hydrogen bomb; the neutrons and X-rays from this blast then activate the fusion secondary to achieve a fusion explosion, which releases energies of higher orders of magnitude. If the plutonium fission primary does not activate, neither will the fusion secondary, and the sealed-pit primary now depended on a complex chain of events unfolding in sequence: symmetrical implosion of the plutonium, activation of the initiator and release of the tritium-deuterium booster.

However, even a partial or low-yield accidental fission explosion would be catastrophic, and so the Sandia weapons designers sought to achieve an added level of protection known as one-point safety. This is where a pit or primary will not go supercritical in the event of an implosion being triggered from a single point in the sphere of high explosive lenses surrounding the plutonium. Supercriticality and the resulting fission explosion can occur only if the implosion is perfectly symmetrical, which in turn depends on the explosive lenses being detonated with perfect synchronization. While it was all too easy to imagine one of the explosive lenses being accidentally detonated – for instance by a bullet or a piece of shrapnel, by debris from a disintegrating plane, or by heat from burning jet fuel – it was considered astronomically unlikely that two or more such events could be triggered at precisely the same time. Thus one-point safety was considered the gold standard for nuclear weapon safety, formalized in 1968 by the assistant to the Secretary of Defense for Atomic Energy, Dr Carl Walske, as having "a probability of less than one in one million of producing a nuclear detonation if a detonation of the high explosives originates from a single point". The original Fat Man bomb had caused such anxiety for its builders precisely because it was not considered one-point safe.

Even with one-point safety, however, there were many ways in which an accident with a nuclear weapon could result in catastrophe. Three of the most notorious Broken Arrow incidents involved sealed-pit weapons that could have detonated, and two of them generated significant levels of radioactive contamination. The least well-known of the trio was perhaps the nearest the world has come to an accidental nuclear explosion: the 1961 Goldsboro incident.

Just before midnight on 23 January 1961, a B-52 bomber from Seymour Johnson Air Force Base, in Goldsboro, North Carolina, was refuelling in mid-air. They were 10 hours into their mission – an airborne alert. Airborne alerts were key to Operation Chrome Dome, the SAC's grand strategic plan for the defence of the USA; by keeping nuclear-armed bombers in the air at all times, American nuclear capability could not be destroyed by an enemy first strike. Thus, the B-52 was armed with two Mk-39 H-bombs, each with a yield of 4 megatons, and tasked with staying aloft for extended periods, ready to head for enemy territory if the balloon went up.

As the boom operator of the refuelling plane observed the B-52 in the darkness, he noticed a trail of pink fluid streaming from its right wing; fuel was leaking out at a shocking rate. In less than two minutes the enormous fuel tank on the right wing was empty. The plane was not far from base and command told Major Walter Tulloch, the pilot, to prepare for an emergency landing. He shut off the engines next to the fuel leak, and then tried to dump the 150,000 litres (40,000 gallons) of fuel in the left-wing tank, in order to balance out the plane. But the left tank would not drain properly, and when Tulloch lowered the flaps to prepare for landing ominous noises ricocheted through the airframe. The enormous bomber went into a barrel roll and started plummeting to earth, spinning uncontrollably. At about 3,000 metres (10,000 feet) there was an explosion, and the plane began to break up in mid-air, starting with the right wing ripping off. Of the eight crew, only five made it to the ground alive.

The Mk-39 bombs were sealed-pit devices, designed to be one-point safe. In order to detonate they had to progress through a series of arming steps. Unfortunately, the catastrophic break-up of the plane unleashed an unlikely confluence of events that triggered almost all of these. A lanyard in the cockpit connected to a manual arming pin in each bomb; one of the safety measures instituted was to make this pin horizontal, so that a bomb accidentally dropping vertically from the plane would not pull it correctly. But, as a 1987 review of the accident from Sandia National Laboratories described: "During the breakup, the aircraft experienced structural distortion and torsion in the weapons bay sufficient to pull the pin from one of the bombs, thus arming the Bisch generator." A Bisch generator is a device that produces pulses of electricity, and the arming of this generator provided the bomb with internal power, enabling subsequent arming steps to proceed.

When the plane broke into pieces both bombs slid free of their bays. As the aft bomb smoothly rolled out, safeing wires running into its body were pulled free, triggering the start of the arming sequence by firing the pulse generator. This in turn switched on the first set of batteries, started a timer and released the drogue parachute, which deployed the main parachute. The altitude-sensing barometric switches closed, and the timer ran out, activating the second set of batteries. The bomb hit the ground in a field near Faro, North Carolina. In the nose of the bomb were piezoelectric crystals, which generate electricity when crushed. As the nose of the bomb crumpled, this crush switch sent a firing signal to the bomb's X-unit, telling it to detonate the high explosives, but it did not go off. Just one step in the safety sequence was all that prevented an apocalyptic calamity. The bomb was equipped with an MC-772 Arm-Safe Switch, aka a ready/safe switch, which was operated from the cockpit. In order for a bomb to be armed, the navigator of the B-52 had to pull out a knob on the control panel and turn a dial from SAFE to ARM, but as the Sandia review reflected, "The nonoperation of the cockpit-controlled ready-safe switch prevented nuclear detonation of the bomb."

If the MC-772 had malfunctioned during the accident and somehow set the bomb to ARM, the Mk-39 would have detonated with more power than all of the explosives used in the Second World War, including the two atom bombs. It would have been a ground detonation, rather than an airburst, and would thus have generated massive amounts of fallout. Nuclear historian Alex Wellerstein calculated the likely result, characterizing it as "a pretty big explosion, with a fallout plume capable of covering tens of thousands of square miles with hazardous levels of contamination (and nearly a thousand square miles with fatal levels)". Depending on wind and weather conditions, such an explosion could have put in mortal danger tens of millions of lives up the eastern seaboard of the USA, and rendered uninhabitable for generations a region that includes Washington, Baltimore, Philadelphia and New York. In his bestselling 2013 book on the story of nuclear weapon safety, *Command and Control*, Eric Schlosser writes that when Defense Secretary Robert McNamara, newly installed in post by the recently elected President Kennedy, heard about the near-miss on his third day in the job, "the story scared the hell out of him".

To modern eyes, the MC-722 is not a device that inspires confidence. It resembles an old-fashioned gas meter, with a heavy die-cast metal rotary dial marked with green and red stripes and the letters A and S. When activated, it sent a 28-volt signal to the arming switch on the bomb. Weapons engineers would soon learn that the MC-722 all too easily malfunctioned, since a simple crossed-wire, or a loose nut or misplaced spanner with the same result, could accidentally generate such a signal. In a 2010 Sandia training film about the Goldsboro incident, Sandia engineers made the following observation: "some people could say, hey, the bomb worked exactly like designed. Others can say, all but one switch operated, and that one switch prevented the nuclear detonation . . . Unfortunately there had been some 30-some incidents where the ready-safe switch was operated inadvertently. We're fortunate that the weapons involved at Goldsboro were not suffering from that same malady." Another Sandia engineer, Parker Jones, wrote in a memo that "one simple, dynamo-technology, low voltage switch stood between the United States and a major catastrophe!" Jones had just investigated a 1962 incident in which a tiny nut had come loose in a control panel in the cockpit of a B-52, creating a short circuit that bypassed the ready/safe switch and arming four Mk-28s, even though the dial in the cockpit hadn't been turned, a mishap only discovered when ground crews came to unload the weapons. If this had happened at Goldsboro, Jones wrote, "it would have been bad news – in spades".

Instead of obliterating North Carolina, the Mk-39 rather comically ended up stuck upright in a field with its parachute draped over it, and the Air Force was able easily to retrieve the radioactive elements of the bomb, as part of the standard nuclear clean-up procedure known as a Moist Mop operation. As for the bomb from the fore bay, the arming switches did not come out in the right sequence, and so its Bisch generator did not arm, its parachute did not open and it plummeted straight down to earth, landing in a field just off Big Daddy's Road, near the Nahunta Swamp. Its high explosives did not detonate but the dense uranium of its secondary had so much inertia that it went through the front of the bomb and penetrated more than 21 metres (70 feet) deep into the boggy ground. An explosive ordnance detail sent to retrieve it ended up digging for a fortnight, scooping out a colossal crater 40 metres (130 feet) wide at the surface and 13 metres (43 feet) deep, before giving up and deciding to leave the secondary where it was. Instead of retrieving it, the Air Force simply bought an easement on a 60 metre (200 foot) circle of land: a legal injunction on ever digging in the area. Other than this forbidden disc of swamp, the other main memorial to the accident is a historic marker in nearby Eureka, which reads: "NUCLEAR MISHAP. B-52 transporting two nuclear bombs crashed Jan.1961. Widespread disaster averted: three crewmen died 3 mi. S."

Despite the unrivalled potential of the Goldsboro incident to have resulted in nuclear catastrophe, and the tragic loss of three airmen, the radiological impact of this Broken Arrow was minor. Perhaps because of this, it remains less well known than two dramatic – and messy – incidents that came after. The first of these, the Palomares incident of January 1966, is probably the most notorious Broken Arrow of all.

Like the aircraft from the Goldsboro incident, the B-52 involved had taken off from Goldsboro on a Chrome Dome airborne alert. Such alerts involved at least three different routes: a western one, which circled around Alaska and the west coast; a northern one, which travelled to Greenland and back (sometimes known as the Thule monitor – see below); and an eastern one, which took bombers across the Atlantic, around Europe and back. This B-52 was on the eastern circuit, and at 10:15 on 17 January 1966, it was high above the coastline of southern Spain, approaching a tanker for a mid-air refuelling. The B-52 came in too fast, crashing into the fuel boom. Flames travelled up the boom into the tanker, which exploded, killing the four-man crew. Meanwhile the B-52 started to break apart. Three of its crew died, but the other four survived. Captain Charles Wendorf landed safely in the ocean, despite his parachute being on fire. The navigator, Captain Ivan Buchanan, ejected through the fireball of the exploding plane, but could not get free of his seat and couldn't get his chute to open. Eventually he freed it by hand and floated to earth but was knocked unconscious when he landed in a field near the small village of Palomares. Co-pilot Lieutenant Michael Rooney had an even more hair-raising escape. He was not in an ejector seat when the accident happened and was flung around inside the disintegrating and cart-wheeling plane until he managed to crawl out of the escape hatch, only for a flaming engine pod to come hurtling past his head, close enough to singe his hair.

The B-52 had been carrying four Mk-28 thermonuclear bombs; it took some time for it to become clear what had happened to them. One had landed just a few hundred metres inland from the beach, not far from Palomares. One of its parachutes had opened, so it had touched down relatively softly and was mostly intact. Two of the bombs had landed less gently, with partial detonation of their high explosives scattering bomb parts and, most alarmingly, plutonium dust. One of them was in the hills and posed no immediate danger to life, but the other had landed near Palomares, narrowly missing a farmhouse, and dispersing plutonium dust onto nearby gardens and tomato fields. The fourth bomb was never located, despite a massive search operation, which initially saw investigators checking hundreds of wells, abandoned mine shafts and other holes where it might be hidden.

The first rule of a Moist Mop operation was to try not to alarm the locals, while the overriding principle of military engagement with the media and the public was blanket denial of any nuclear involvement or risk. Both these principles would be tested to destruction by the public relations storm around the Palomares incident, which saw the world's press descend on the tiny village, so poor and undeveloped that it did not have running water and its

Handle with care

The plutonium core for the Trinity Gadget is loaded into an Army sedan, for transport from the McDonald Ranch to the Trinity test site, where the final assembly will take place. Harry Daghlian is the right-hand man of the two carrying the core.

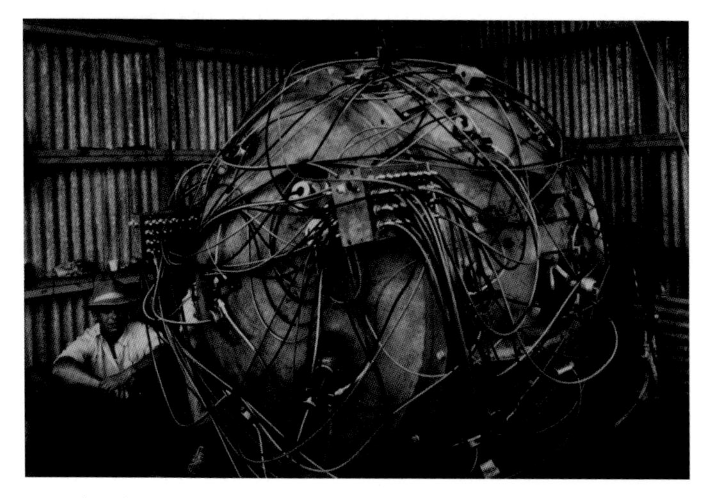

Doomsday device

The Gadget, festooned with cables connecting the X-unit detonators to the high explosives, sits in the shed atop the tower at the Trinity test site.

Fireball

This rare colour still captures the expanding fireball just a few hundredths of second after detonation of the Trinity test device on 16 July, 1945.

Playing it cool

Louis Slotin (left) and Herb Lehr help to assemble the Trinity Gadget in the tent at the base of the shot tower, prior to the Trinity test. Slotin is looking at the 'slug', a heavy cylinder of uranium, inside which sits the plutonium core.

The Demon Core

Re-enactment of Louis Slotin's fatal demonstration with the Demon Core, showing how he would have held the upper hemisphere of tamper in a bowling ball grip with his left hand, while using the screwdriver blade to keep a minute gap between the two hemispheres.

Loading the X-10

Dummies help to demonstrate how workers would have loaded uranium fuel into the loading face of the X-10 Graphite Reactor, where intense neutron bombardment would transmute some of it into plutonium.

The most polluted place on Earth

A sign warns unwary travellers to stay out of regions still contaminated from the Kyshtym disaster of 1957, when a tank of radioactive waste exploded at the nuclear plant in the secret city of Chelyabinsk-40 (known today as Ozersk).

Cows and Cockroft's Follies

Cows grazing peacefully next to the Windscale nuclear plant, just a year after the disastrous fire of 1957. Visible behind them are the chimneys crowned with Cockroft's Follies, distinctive filter galleries that helped limit the escape of radioactive particles.

Broken Arrow

The shattered fuselage of the downed B-52 near Palomares in southern Spain, in January 1966. The four thermonuclear bombs it had been carrying were scattered across land and sea.

Up from the deep

Recovery of the 'Broken Arrow' lost thermonuclear device, from the seabed off Palomares, afforded the public their first ever look at an American H-bomb.

Presidential briefing

President and Mrs Carter visit the control room of Three Mile Island Unit 2 on 1 April, 1979, accompanied by Pennsylvania's governor, Dick Thornburgh (second from left), and Harold Denton of the Nuclear Regulatory Commission (far left).

Three Mile Island

Aerial view of Three Mile Island, in the Susquehanna River in Pennsylvania. The two cylindrical buildings house the reactors; Unit 2, which suffered the meltdown in 1979, is the one nearest the top of the picture.

All clear

An operative in full protective gear emerges from an airlock at Three Mile Island, to give the all clear for the room beyond, as the crew participates in the ongoing clean-up effort following the 1979 meltdown.

Ghost town

A view over the abandoned city of Pripyat, towards the Chernobyl power plant, showing the iconic venting stack, the dilapidated original 'sarcophagus' containment structure, and the newer containment arch, prior to being moved into place.

The Elephant's Foot

An intrepid and probably foolhardy photographer snaps a close-up of the Elephant's Foot, a highly radioactive formation of solidified corium lava, in the bowels of the destroyed Unit 4 reactor building.

Chernobyl today

The ruined control room of Unit 4 of the Chernobyl nuclear power plant, which, along with the rest of Unit 4, is now entombed within a colossal arched containment sarcophagus.

Blown apart

Aerial view of the ruined Unit 4 reactor building, just a few days after the reactor exploded. The remains of the core were still smouldering within the building, spewing radiation into the sky.

Mementoes of a catastrophe
Ghoulish remnants adorn the shattered halls of the abandoned city of Pripyat, once home to 50,000 Chernobyl nuclear power plant workers and their families.

Disaster in progress
A satellite view of Fukushima Daiichi power plant, just after the second reactor explosion on 14 March, 2011. The waters of the tsunami have receded but the plant is in the midst of a crisis.

Exclusion zone
A dog picks its way through tsunami debris that still litters the streets of Tomioka town, within the exclusion zone around Fukushima Daiichi, a month after the problems at the power plant.

inhabitants could muster only two cars and a single phone between them. Residents were alarmed by the 1,700 US military personnel combing the region, many of them brandishing Geiger counters, although these were ill-suited to the task because the plutonium for which they were searching emitted alpha radiation, easily blocked by even a blade of grass, and thus fiendishly hard to detect.

The US Air Force initially insisted that the plane was carrying only "unarmed nuclear armaments" and failed to mention that one of the bombs was missing. "There is no danger to public health or safety as a result of this accident," they claimed. After weeks of worsening headlines, they admitted that a nuclear weapon was missing – probably at sea, in a synchronous echo of the plot of the previous year's James Bond film *Thunderball.* Headlines blared gleeful Soviet accusations that the missing weapon constituted a "'Nuclear Volcano' in Sea off Spain". But the USAF continued to cover up the most disturbing aspects of the incident, prompting the Spanish Nuclear Energy Board, presumably acting on the assurances of the US Government, to insist that "there is not the slightest risk in eating meat, fish, vegetables from the [affected] zone, or of drinking milk from there."

The limited plutonium-tracing capability of the searchers detected contamination of a 1.6 kilometre/1 mile-long strip that extended through the village of Palomares and included surrounding tomato fields. Despite this, the villagers were not evacuated and the area was not cordoned off, because, according to a later report from the Defense Nuclear Agency, of "the politics of the situation". Nonetheless, the US Government did promise to decontaminate the area, and to this end 4,000 truckloads of produce was harvested and burned, and 850 cubic metres (30,000 cubic feet) of soil, weighing about 1,500 tons, was packed into barrels and shipped to the US military nuclear waste facility at Savannah River, North Carolina. More soil and harvested tomatoes were buried around Palomares, while about 40 hectares (100 acres) of land around the bomb sites was eventually fenced off. The only pieces of protective equipment offered to the soldiers tasked with digging up the soil were surgical masks, even though the military was fully aware that they had only placebo value.

In truth the Americans had no idea whether they had successfully cleared up all the plutonium, and they never revealed how much had actually been spread around, although a likely estimate is about 2.7 kilograms (6 pounds). Mystery also surrounded the fourth bomb. Analysis of wind conditions, the discovery of a piece of the bomb's tail on the beach, and a report from a local fishermen of seeing a "stout man" parachuting into the sea, combined to suggest it was somewhere on the sea bed. A huge search operation was launched, involving ships, aircraft, over a hundred divers and four manned submersibles, all under the eyes of loitering Soviet vessels. Rear Admiral William Guest, in charge of the operation, pointed out, "It isn't like looking for a needle in a haystack, it's like looking for the eye of a needle in a field full of haystacks in the dark." Eventually the drowned bomb was located by one of the submersibles, and, after two attempts, was recovered. The Navy proudly displayed their find to a group of journalists invited aboard Guest's flagship; it was the first time an American thermonuclear weapon had been shown to the public.

At the time of the accident, Spain was a dictatorship ruled by General Franco; no one from Palomares was likely to make a fuss when the party line was to follow the Americans' lead. Fifty years later, however, the situation was different, and a European Union (EU) investigation uncovered disturbing evidence that the Moist Mop clean-up operation, and subsequent Spanish follow-up, had been woefully inadequate. Nobody had prevented local people from picking up pieces of debris to keep as souvenirs, let alone checked whether the souvenirs were radioactive. Rabbit hunters had broken down the fences meant to keep people out of the prohibited zone, but no one knew if the meat they were eating was contaminated. Although, at 5 millisieverts (500 millirem) a year, radiation dose levels measured around Palomares were only about twice the background level, the EU decreed that another 57,000 cubic metres (2 million cubic feet) of soil needed decontamination or removal. In 2015 America and Spain agreed to put in place a $35 million, three-year programme to decontaminate soil and perform blood tests on the residents of Palomares.

In the wake of the Palomares incident the Pentagon reassessed its commitment to Operation Chrome Dome and the expensive and accident-prone airborne alerts. But the power of the SAC would not be denied; its influential generals insisted that the alerts were vital to national security. The administration of Lyndon Johnson compromised by reducing the number of airborne alerts running per day to just four. Among the four missions deemed worthy of continuing was Operation Hard Head, aka the Thule monitor: a flight pattern that had a B-52 describing a giant figure of eight at an altitude of 10,700 metres (35,000 feet) over a crucial US radar installation at Thule in western Greenland. This base was a lynchpin of the Ballistic Missile Early Warning System (BMEWS), a set of radar installations intended to detect enemy missile launches quickly enough to give the USA a few vital minutes early warning of an imminent missile strike. One of the nagging fears of the US military was that a strike on the Thule base – perhaps using conventional weapons deployed by enemy special forces – could disable this vital element of the nation's early warning system, leaving the USA blind to missiles approaching from the Arctic north-east. Accordingly the Thule monitor was one of the more valued airborne alerts: a B-52 circling over Thule, combining its bomb-carrying deterrent mission with a straightforward visual check-up to ensure that the radar installation was still there.

On 21 January 1968, the Thule monitor B-52 caught fire at its cruising altitude after one of the co-pilots stuffed foam-rubber cushions under a jump seat to try to get more comfortable. What he hadn't realized was that under the seat was the outlet for a heating vent, and when, about five hours into the flight, the heating was switched on, the cushions were set alight. Attempts to fight the fire had no success, and as the pilot radioed the air base at Thule to request an emergency landing, the cockpit filled with smoke. At about 16:37 local time the plane lost electrical power and the pilot, Captain John Haug, told the crew to prepare to bailout.

It was deep midwinter in Greenland, dark and brutally cold. Haug guided the stricken plane to within a few miles of the airbase, so that they could bail out as close to rescuers as possible. He was the last man to bail out, about 6 kilometres (4 miles) short of the runway. Of the seven-strong crew, six managed to bail out successfully, but one man, Captain Leonard Svitenko, had smashed his head as he exited the aircraft and never regained consciousness. The other six reached the ground in varying states of distress. Major Alfred D'Amario and Captain Haug landed on the base itself and were able to walk into nearby hangars, while navigator Captain Curtis Griss (known in some sources as Criss), had dislocated his shoulder during the ejection and landed 10 kilometres (6 miles) away from the base. He was not rescued for almost 24 hours, and frostbite claimed both his feet.

The plane itself flew on, passing over the base and describing a wide arc to port, over the Wolstenholme Fjord, losing altitude all the time. Finally, it smashed into the ice sheet over Bylot Sound at the mouth of the fjord, around

11 kilometres (7 miles) west of Thule, leaving a streak of debris about 4 kilometres (2½ miles) long. The high explosive shells detonated in all four Mk-28 bombs, but the one-point safety safeguards all operated successfully, and there was no nuclear yield. This was particularly fortunate because if the USA had detected a nuclear explosion at Thule, it could well have misinterpreted it as evidence of a Soviet first strike, triggering retaliation and starting a war. However, plutonium and uranium from the bombs had been scattered across an area of about 8 square kilometres (3 square miles), while radioactive dust had spread for miles with the smoke from the burning wreckage. Around 6 kilograms (13 pounds) of plutonium was dispersed.

This mess was located on Danish, rather than US, territory, 1,100 kilometres (684 miles) north of the Arctic Circle in the perpetual dark of winter and amid temperatures as low as –60°C (–76°F) and winds of up to 143 kilometres per hour (89 miles per hour). Cleaning it up would be an unforgiving task, and the Americans were minded to simply dump the wreckage into the fjord. But the Danes objected, and since the vital US base was on their territory, the Americans conceded, launching a clean-up operation that was officially called Project Crested Ice, but unofficially known as "Dr Freezelove", after the 1964 film about SAC shenanigans, *Dr Strangelove*. The contaminated area was delineated by a Hot Line, and the top 5 centimetres (2 inches) of ice within the Hot Line – some 6,700 cubic metres (237,000 cubic feet) of material – was removed, taken back to the base, compacted, packed in containers and shipped back to the USA, via ship and rail, to an Atomic Energy Commission facility in Aiken, South Carolina. The material shipped back filled 147 freight cars. What the US Government did not tell the Danes, however, was that the crash had punched a hole through the sea ice, and some of the wreckage, including at least part of one of the thermonuclear devices, had fallen through into the water beneath, to settle on the floor of Bylot Sound or be carried away by the current.

When the ice thawed in the summer, the US Navy sent a mini submarine to look for the missing material (probably the uranium and lithium deuteride "spark plug" from the fusion secondary of one of the bombs), but care was taken not to alarm the Danish Government. "For discussion with the Danes," commanded a classified Pentagon memo uncovered by a BBC investigation 40 years later, "this operation should be referred to as a survey – repeat survey – of bottom under impact point". The missing material was never recovered, and in September the mission was wound up.

The primary consequence of the Thule crash was to put an end to SAC's Chrome Dome airborne alert programme, the day after the accident. It was pointed out that if the B-52 had crashed into the base, or the bombs had gone off, a misplaced seat cushion could have triggered World War Three. The Thule Monitor mission continued, however, with a B-52 circling high over Greenland every minute of every day, but without carrying thermonuclear weapons.

The accidents described in this chapter represent only a fraction of known American Broken Arrows. Less is known of Broken Arrows among other NATO members, while the records of equivalent incidents among the nations of NATO's opposition, the Warsaw Pact, are a closed book. However, given the lower safety standards in the Soviet sphere, and the worse levels of training, equipment, technology, maintenance and accountability in comparison with the Americans, it would be stretching credulity to imagine that the annals of Soviet accidents do not include incidents at least as hair-raising as Goldboro or Thule. That the world has never experienced an accidental Nucflash is partly testament to the hard work and expertise of weapons safety engineers, but also, as the cases above illustrate, partly down to sheer luck.

Chapter 8

HUMAN ERROR: THREE MILE ISLAND, 1979

Alarms blare and panicked voices are raised in the control room as operators race to manage water levels in a reactor spinning out of control. Faulty instruments mistakenly cause the operators to believe there is too much coolant being pumped into the core, when in fact the opposite is happening. They frantically dump water from the reactor just when they should be adding it. The core careers ever closer to the doomsday scenario, in which the fuel rods overheat and fuse into a molten mass of increasingly hot radioactive elements, which will burn through the bottom of the reactor vessel, through the floor of the containment building, and perhaps through the Earth itself . . . all the way to China.

This was the opening scene of the thrilling new film *The China Syndrome*, starring Jack Lemmon, Jane Fonda and Michael Douglas, which hit US cinemas on 16 March

1979. Less than a fortnight later almost precisely the same scenario would be played out for real, in the control room of a genuine nuclear power plant, and millions of people would be gripped with fear about the possibility of a real-life China Syndrome that might irredeemably contaminate one of the most densely populated regions of the planet for thousands of years to come.

In the film, a television news crew is serendipitously present to witness – and furtively record – the dangerous incident. In real life, one of the first people outside the nuclear industry to know that something was afoot was Dick Thornburgh, the recently elected governor of Pennsylvania. It was at 07:50 on the morning of Wednesday, 28 March 1979, in the middle of a morning meeting with state legislators, that Governor Thornburgh got the call. The state director of Emergency Management informed him that there had been an accident at the Three Mile Island (TMI) nuclear power plant, located just 16 kilometres (10 miles) downstream of his office, in the middle of the Susquehanna River. "Nuclear jargon was a foreign language to me," Governor Thornburgh reflected 20 years later. "I knew enough, however, that the thought of issuing a general evacuation order first entered my mind at 7:50 that morning and never left me through the unprecedented days of decision that followed."

Two days later, on Friday, 30 March, the governor got an even more alarming call. The voice at the other end of the line told him that a fog of radioactive iodine might have been released from the stricken reactor and might even now be stealing into the homes of the people of Harrisburg and into the mouths and lungs of children and pregnant women. Once ingested, this pernicious substance would quickly concentrate in infant thyroid glands, working insidiously to trigger a plague of cancers. Saturday brought the most frightening call yet: the crippled reactor might be about to explode, showering deadly radioactive contamination over two million people. It might be necessary, he was told, to order an immediate evacuation of everyone within a 16–32 kilometre/10–20 mile-radius. The day after that, the President came to visit.

This was the extraordinary sequence of events that unfolded after the worst civil nuclear disaster in American history, the meltdown at Three Mile Island. In the event, the reactor did not explode and there was no significant escape of iodine-131, nor of any other dangerous nuclides, but there was a catastrophic core meltdown, a massive public panic and a lasting legacy of fear and distrust around nuclear power and the nuclear industry. Some of that distrust resulted when it emerged that many of the design flaws and problems with the TMI reactor were, or should have been, known about by the nuclear industry and, specifically, the plant's operators. For it was no accident that the plot of *The China Syndrome* proved uncannily prescient; almost every aspect of the TMI incident could have been foretold, for almost every aspect had already been played out.

The American civil nuclear power programme had its start in President Eisenhower's "Atoms for Peace" initiative, an attempt to rebrand nuclear science away from its military overtones. In 1958, Eisenhower opened the United States' first non-military nuclear reactor at Shippingport in Ohio. But this reactor was essentially a scaled-up version of the reactors that the US Navy had developed for its nuclear-powered submarines. These were the result of a formidable research and development campaign by Hyman Rickover, initially a captain and later an admiral, the man in charge of the Navy's top-secret nuclear submarine programme.

Rickover had driven through the development of a stunningly successful, small nuclear power plant that would revolutionize submarine warfare, through a combination of ambition, brilliance, intense focus and even a willingness to disregard federal law. One of the technical challenges Rickover had tackled was the need to develop an entirely new metal alloy to use in cladding the uranium fuel rods at the heart of his reactor design, an alloy capable of withstanding brutally high temperatures without melting or corroding. Rickover had identified the high-melting point, corrosion-resistant metal zirconium as the best candidate, and developed a high-performance alloy known as zircaloy to use for fuel cladding. In order to test the resilience of the new alloy-clad fuel elements, however, he needed to expose them to the heavy radiation environment of a high-power research reactor. The problem was that the only suitable one was the NRX, Canada's National Research Experimental reactor, which began operating in 1947 and was for many years the world's most powerful. Rickover sweet-talked officials at the NRX into giving him test time, conveniently ignoring the fact that his own nation's federal law prohibited the transport of enriched uranium out of the USA. Rickover simply had a model fuel assembly labelled "materials test" and shipped it to Canada anyway.

An interesting development that would later prove relevant to the TMI incident was that shortly after this, in 1952, there was an accident at the NRX and Rickover volunteered US military personnel to help with the cleanup. Leading the contingent from the US Navy was a young officer, James Earl Carter, who in 1979 would be President of the United States. The time he spent toiling to scrub clean floors and walls, while encased in a heavy rubber suit and respirator, is said to have affected his opinion of nuclear reactors thereafter.

The reactor that Rickover designed, and which became the basis for the workhorse of the US civil nuclear reactor fleet, was the Pressurized Water Reactor (PWR), an ingenious design in which water is used as moderator, coolant and energy transfer medium. Water, under very high pressure, fills the core vessel and bathes the zircaloy clad fuel rods, picking up heat and circulating to a heat exchanger, known as the steam generator, where a secondary loop of coolant absorbs the heat and turns into steam in order to drive steam turbines. The water in the primary loop is kept under pressure so that it remains liquid even at high temperatures. This design has many advantages, including the ability to keep the primary coolant, which bathes the fuel elements and thus picks up radioactive contamination, sealed away and separate from the secondary circuit. PWRs can use relatively low-grade enriched uranium for fuel and do not need plutonium, which makes them safer since the fuel is much less likely to be able to achieve a critical configuration in case of an accident. Perhaps most crucially, the water in the core acts as a self-regulating moderator for the fission of the fuel: the more intense the fission, the hotter the water gets; and the hotter it gets, the less dense it becomes. The less dense the moderator, the fewer neutrons are moderated and the lower the likelihood of neutron capture and subsequent fission. Thus fission in a PWR becomes self-limiting.

But PWRs also have many inherent limitations and vulnerabilities, several of which would be exposed in the TMI incident. They depend on complex, highly engineered systems of water and steam circulation, with many pumps and valves. A key element of the PWR is the pressurizer, an upright cylindrical vessel separate from the reactor vessel but connected to it by a pipe and thus part of the same primary coolant system. Pressure regulation in the pressurizer is achieved by raising or lowering the temperature of fluid in the vessel, since water pressure depends on its temperature. The pressurizer regulates the pressure throughout the entire primary system, ensuring that the coolant never boils and always remains in a liquid state, which is essential for the operation, and indeed the survival, of the core. If the core is not covered with liquid coolant, the enormous amount of heat generated by fission will quickly cause the fuel elements to heat up to melting point, causing a meltdown.

In the relatively small reactors used in submarines, such core meltdowns were of less concern than the possibility that pressure changes – within a closed, high-pressure system – might cause shockwaves, known as water hammer, which could easily break apart vulnerable plumbing. To guard against this possibility, a key feature of the pressurizer was the presence at its head of a bubble of steam. Liquid water is incompressible, which is why water hammer is such a threat, but steam is compressible. The steam bubble thus acts as a sort of damper or buffer, like a shock-absorbing spring. Any sudden changes in water pressure thus simply compress or expand the steam bubble, instead of causing water hammer. If the pressurizer fills up with water, however, then there is no steam bubble; this is known as "going hard" or, more commonly in the Navy and the nuclear industry, "going solid". Letting the pressurizer "go solid" was *the* cardinal sin of PWR management on a submarine, because it brought instant, massive risk of water hammer, with concomitant catastrophic leaks and/or steam explosions, which must inevitably have fatal consequences in the confines of a submarine. Naval reactor engineers were thus trained above all to manage the pressurizer and avoid going solid, a tendency that would prove to be extremely important in the story of TMI.

Rickover's success made PWRs the standard for submarine fleets around the world, and meant that, in the competition between reactor designs to see which would succeed in the civil market, PWRs were at the forefront. Shippingport was the first civil PWR; many more were to follow. By the late 1970s about 80 per cent of the reactors being operated or under construction in the USA were PWRs. Making them safe enough for giant power plants, sometimes located in or near population centres, meant instituting a series of elaborate and complex safety mechanisms.

One of these is a valve on top of the pressurizer, called the pilot-operated relief valve (PORV), which opens in case of over-pressure events that might otherwise burst the system, relieving pressure by venting steam and superheated water. Since this will be coolant from the primary system, it will be contaminated, and hence is allowed to vent only into the interior of the containment building. In a PWR, maintaining the relatively low operating temperature of the primary coolant (which is not supposed to go above 315°C/599°F) depends on the successful operation of the steam generator, which transfers energy to the secondary coolant loop at a tremendous rate. If this transfer is interrupted for some reason, the temperature and thus the pressure of the primary coolant will rocket, and so PWRs are equipped with a complex set of devices to cool the reactor in case of emergency. Collectively they are known as the Emergency Core Cooling System (ECCS), and they can circulate huge amounts of coolant around the primary system if needed.

At the start of the Atoms for Peace era, the hopes invested in civil nuclear energy were positively utopian, epitomized by the famous boast of Lewis Strauss, Chairman of the Atomic Energy Commission (AEC), that nuclear energy would be "too cheap to meter". By the 1970s, that fervour had somewhat diminished and Strauss's prediction had fallen far short of the mark, not least because of the expense of adaptations such as the ECCS. The financial burden of constructing a nuclear power plant was matched only by the regulatory burden. Accordingly, it was generally a lot easier to install a second reactor at an existing plant than to start a new plant. This was the plan at the Three Mile Island nuclear power plant, which had been constructed on the eponymous island in the Susquehanna River, about 16 kilometres (10 miles) south of Harrisburg, the state capital of Pennsylvania, and 3 miles (5 kilometres) downriver from Middletown (hence the name).

Metropolitan Edison (Met Ed), a subsidiary of General Public Utilities, began construction of the first reactor, Three Mile Island 1 (TMI-1), at the north end of the island in 1968. It was completed and commissioned in 1974, began generating electricity and continued to do so for 45 years, finally closing in September 2019. TMI-1 was one of the great success stories of the American civil nuclear power industry. Its smooth operational history fit well with the prevailing narrative of the nuclear industry, which had worked hard for over two decades to protect its reputation and to gloss over anything that might trigger public concern. Twenty years after the disaster, Governor Thornburgh would recall that his beliefs reflected the success of this strategy: "Nuclear power was still the technological marvel of our time . . . an industry whose safety record had been, or at least was thought to have been, second to none." But in March 1979 Thornburgh himself already knew that this was not completely true, because, as he would soon have cause to recollect, he had read the anti-nuclear book *We Almost Lost Detroit*.

John G. Fuller's 1975 book concerned the 1966 Fermi-1 meltdown, an incident at Unit 1 of the Enrico Fermi Atomic Power Plant in Newport, Michigan, about halfway between Detroit, Michigan, and Toledo, Ohio. Fermi-1 was a fast breeder reactor: a nuclear reactor that produces more radioactive fuel than it consumes. Uranium fuel is typically comprised mostly of non-fissile U-238, which can be transmuted into fissile nuclides through capture of neutrons. In the 1950s uranium was in relatively scant supply because little investment had yet been made in locating and mining reserves, and so it was believed that availability of uranium would be the limiting factor for nuclear power generation. Breeder reactors offered a solution to this problem: if enough of the right type of neutrons produced in fission can be retained in the fuel element, they can be captured by non-fissile nuclides, transmuting them into fissile ones.

Theoretically, a breeder reactor can make use of almost 100 per cent of its fuel, leaving virtually no waste and extracting every last joule of energy. In order to start up such a reactor, however, an expensive blend of uranium and plutonium is required, and eventually breeder reactors would fail to prove economically viable, but in the late 1950s, when Fermi-1 was designed, breeder reactors looked like the future.

Fermi-1 used molten sodium metal as a coolant. Liquid metal coolant offered many advantages, such as high energy density and ease of heat transfer, but such materials bring great challenges. Sodium is highly reactive, corrosive and toxic. Any leak or dispersal of such material is a major problem.

Fermi-1 first went critical in 1963, and after years of low-power testing it moved into a new phase of high-power testing at the end of 1965. Testing runs in August 1966 produced anomalous readings, with suspiciously high temperatures recorded in some parts of the reactor. The reactor was shut down and rejigged to allow another testing run to begin on 4 October. On 5 October, at 15:09, alarms went off to indicate abnormal levels of radiation in exhaust vents, and the reactor was shut down again. It took almost a year to locate the cause of the incident, which was discovered only when investigators managed to lower a periscope to the bottom of the reactor, and worked out that the free circulation of liquid sodium coolant had been interrupted when one of the coolant channels became blocked by a thin disc of metal; this was widely reported as having been a squashed beer can, left inside the reactor by a careless worker. Several months more passed before workers were able to extract the disc, in an operation akin, Fuller wrote, to "taking out an appendix through the nostrils". They were able to ascertain that there were, in fact, two discs, and that they were not beer cans but the zirconium liners from "flow guides", which had previously been welded to the bottom of the coolant channel precisely in order to prevent catastrophic consequences from a fuel assembly meltdown. Unfortunately, the flow guides were late and poorly thought-through additions to the design and proved unable to cope with the rigours of a liquid metal nuclear reactor. Two of the zirconium liners applied to the guides as heat-proofing had come loose, been swept upwards by the flow of coolant, and plastered over the bottom openings of some of the channels that admitted coolant to the reactor core. Some of the fuel sub-assemblies then overheated and melted.

In his book, Fuller conjured a near-miss doomsday scenario, in which a core meltdown at Fermi-1 came perilously close to resulting in molten uranium and plutonium pooling at the base of a reactor to trigger a critical excursion, massive explosion and catastrophic dirty-bomb-style contamination of a vast metropolitan region. The book drew its alarming title from a quote Fuller attributed to an anonymous engineer investigating the incident, who supposedly remarked, "Let's face it, we almost lost Detroit." Fuller had a distinct anti-nuclear agenda, and his central thesis was categorically rejected by the plant's owners, Detroit Edison (DTE), who went so far as to issue a rebuttal snappily titled *We Did Not Almost Lose Detroit*. They pointed out that less than 1 per cent of the fuel had melted (about 18 kilograms/40 pounds), there was never any danger of it escaping the containment vessel, and there had been no risk of a critical excursion. A DTE spokesman, Guy Cerello, insisted that the incident never tested the boundaries of the plant's safety envelope, and that the book's claims were highly exaggerated: "We did not lose Detroit and we did not almost lose it, either."

While this may have been true, there were several aspects of the incident that should have raised alarm bells. For one thing, the incident had many important similarities with a 1959 incident at the Sodium Reactor Experiment (SRE) in California. The SRE was a small, experimental liquid metal cooled reactor, in which 13 of the core's 43 fuel elements had melted after problems developed with heat transfer by the coolant. In both cases, the operators of the reactors were presented, through the readings given by their instruments, with the evidence of dangerous overheating of the fuel assemblies, and in both cases the operators assumed that the instruments were faulty rather than the reactor. At both the SRE and Fermi-1, according to an analysis by the Union of Concerned Scientists, "workers . . . failed to recognize numerous warning signs that inadequate cooling was damaging the reactor core". Rather than heeding the warning signs and acting to limit or halt the damage, operators took steps that actually made it worse. This kind of operator-exacerbated error would play a key role in the TMI incident, as would the failure to evaluate without preconceptions the story the instruments were presenting. Perhaps more disturbing, however, was the evidence of systematic failures in the way the nuclear industry responded to accidents. Failure to learn from past mistakes meant that the control room operators of the future were doomed to repeat them.

The smooth running of TMI-1 meant that it was largely free of such concerns, but TMI-2 proved to be ill-starred from early in its life. Construction on the second reactor began in 1969, at a site just to the south of TMI-1. Like its sibling, TMI-2 was a Babcock & Wilcox (B&W) model 177FA pressurized water reactor. Despite starting just a year later, construction delays and operational difficulties meant that TMI-2 did not come on line until April 1978, four years after TMI-1 began feeding electricity to the grid. Even then, problems abounded and shutdowns were frequent, and it later transpired that the reactor's operators had falsified operational data so they did not have to keep submitting reports to the Nuclear Regulatory Commission (NRC), which in turn would have led to the Commission imposing further interruptions.

The basic design of TMI-2 was similar to most other civil PWRs. The reactor vessel was a round-bottomed cylinder, plugged with the control rod assembly, which raised and lowered the control rods. The highly pressurized water coolant circulated from the reactor to the steam generator, then through a pump that passed it back into the reactor. A side branch of the primary system led to the pressurizer, which regulated the pressure of the primary system, and was in turn connected to relief valves such as the PORV. All of this equipment sat within the reactor containment building, so that a vent, leak or spillage of primary coolant would not escape into the outside world.

A secondary circulation of coolant ran to the steam generator, where it picked up heat from the coolant circulated from the reactor conducted through the walls of thin metal tubes which ran in loops through the generator and expanded into steam. The steam passed to the turbine, which in turn was connected to the generator, producing electricity for the main grid. From the turbine the secondary coolant passed through a condenser and out of the turbine buildings to the giant cooling towers, to shed heat, before passing back to the condenser and then through a series of pumps to be driven back into the stream generator. The turbine, generator, condenser and pumps were housed in the turbine building, next to the reactor containment building. The final part of the puzzle was a series of eight enormous filtration tanks, where the thousands of tonnes of water that circulated from the turbines to the cooling towers were cleaned by passing them through masses of resin beads.

It was in one of these tanks that the TMI incident began. Lines carrying compressed air had been run to these tanks so that the resin beads within them – which tended to clump together – could be loosened and separated with jets of compressed air. On the afternoon of 27 March 1979, a problem with the resin in Tank 7 had seen operators trying exactly this, but someone had forgotten to close a valve, and for the next 10 hours water had been backing up one of the air lines. If it reached the head of the line it would trip the valves on top of all eight tanks, bringing the secondary circulation to a screeching halt. The electrical backup system that was supposed to guard against this incredibly obscure and unlikely event had not been wired up. Bill Zewe, the shift supervisor for TMI-2 in the early hours of 28 March, had been down at Tank 7 checking on efforts to loosen the resin, and some time before 04:00 he returned to the control room and asked Fred Schiemann, the foreman, to go down and supervise.

At Tank 7, Schiemann climbed up on top of a huge water pipe to look down through the sight glass into the tank. The pipe rumbled reassuringly beneath his feet, but then, at 03:58, it went ominously quiet. Pressurized water had backed far enough up the air line to shut down the whole secondary system, an abrupt change that would inevitably trigger a terrifyingly powerful water hammer: a shock wave of pressure that Schiemann could feel hurtling up the pipe beneath him. He later described it as "loud thunderous noises, like a couple of freight trains". Schiemann managed to jump clear just in time, as the pipe ripped out of its mounts, tearing clear of the pump, which flew into pieces. Scalding hot water blasted into the room.

A panel full of alarm tiles lit up in the control room and the warning horn began sounding. There were over a hundred flashing lights, although the warning signals did not distinguish between serious and trivial issues, moving one of the operators, Craig Faust, to later observe, "I would have liked to have thrown away the alarm panel; it wasn't giving us any useful information". Automatic systems sprang into action: with no steam going to the turbine, sensors tripped, taking it off-line. With no heat being removed from the primary water system, temperature and pressure in the reactor would rocket; accordingly, eight seconds after the turbine cut off, the automatic scram kicked in, throwing control rods into the reactor to soak up neutrons and shut down all fission. Three seconds later, as the pressure in the primary system rose, the PORV opened, relieving the pressure by venting boiling water and steam into the drain tank in the containment building. The OPEN signal that had been sent to the PORV valve registered as a red light on the control console. The pressure in the primary system, which had spiked initially, quickly fell back below danger levels, and 10 seconds after it had opened, the red light on the control room panel went out, apparently indicating that that the PORV was no longer receiving an OPEN signal and must therefore, by default, be closed. To the men in the control room, the system seemed to be working as designed, and as normal in the event of a turbine trip – a reasonably routine occurrence.

But then things started to go screwy. At 04:02 the high-pressure injection (HPI) pumps, one of the main components of the ECCS, automatically switched on. At the same time, the instruments in the control room seemed to be giving contrasting readings: the pressure in the primary system was dropping, but the temperature was climbing – normally the two were coupled together. The operators could not understand why the ECCS had come on, but what worried them more was that it looked as though as the pressurizer was filling up with water and might go solid. By this time Schiemann, having sprinted up eight flights of stairs, was back in the control room; alarmed at the readings from the pressurizer, he ordered one of the HPI pumps be turned off, and the other turned down.

Still the pressure in the system was dropping, the temperature was climbing, and the pressurizer appeared to be filling. Anxious that it was about to go solid, the operators turned off the second HPI pump, but it didn't seem to help. At this point Bill Zewe had a flash of insight that could have saved the situation. What if the PORV had not actually shut, despite the light on the control panel going out? That would explain why the pressure in the system was dropping. One way to tell would be from the temperature reading of the PORV outlet: if it were high, it would indicate that superheated steam was venting; in other words, that the valve must be open. Zewe asked one of the other operators to call out the PORV outlet temperature reading; "228°[F]" came the answer. This did not seem overly high, but unfortunately in the panic the operator had read the wrong readout. The actual reading was 283°F (139°C), a

55° discrepancy that would have revealed the explanation for everything that was going wrong. For although the light on the board indicated that no OPEN signal was being sent to the PORV, it had not closed; the PORV was stuck open, and thousands of litres of primary coolant were boiling out of the reactor, up through the pressurizer and out into a tank in the containment building.

With the PORV open there was a very real danger that the reactor could boil dry, leaving the reactor fuel uncovered. This was why the ECCS had turned on the HPI, in an attempt to replace the escaping coolant, but the operators had turned it off for fear of overfilling the pressurizer. In fact, the pressurizer probably only looked as though it were going solid because the coolant within was boiling so violently. What the operators needed to do to save the reactor was to shut the block valve, a back-up valve that could close off the PORV, seal the primary system and prevent the coolant from escaping.

Should the operators have given more serious consideration to the possibility that the PORV was at the root of the problem? There is a good case to be made that they should have been alert to the likelihood of a sticky PORV, because almost exactly the same thing had happened in 1977 at the Davis-Besse plant in Oak Harbor, Ohio, in an accident with many similarities to TMI. In this incident a pump failure in the secondary system had knock-on effects on water pressure in the reactor, but operators were baffled when they were unable to bring this pressure under control, even after emergency shut down. Fortunately the shift supervisor noticed in time that the PORV was stuck open (it had smashed itself to pieces trying to open and shut because an electrician had "borrowed" and never replaced one of its plug-in relays so that he could patch up a circuit he was repairing), and was able to shut the back-up "block valve", and thus prevent steam escaping from the reactor.

Before this happened, however, the control room engineers had made things worse by overriding the HPI automatic safety systems that injected water into the core for emergency cooling, because the pressure readings convinced them that the pressurizer was in imminent danger of going solid. They were too fixated on an illusory threat to the pressurizer to consider the more important issue: the threat of the core becoming uncovered. Because the reactor had only been in a testing, start-up mode, it had been running at 9 per cent power, and so was not in danger of a meltdown. If it had been running at just 50 per cent power, however, it would have experienced a catastrophic meltdown. The warning signs were there, but neither the NRC nor the reactor manufacturers, B&W, passed on this potentially vital information to other owner-operators of B&W PWRs, such as TMI-2.

In addition, it was significant that both the shift supervisor on duty at Three Mile Island in the early hours of 28 March, Bill Zewe, and the foreman, Fred Schiemann, along with at least two of the control room operators, had trained in the Navy. One of the issues highlighted by the

Davis-Besse incident was the influence of a naval background on control room engineers. Navy engineers who had trained in the operation of submarine PWRs, tending their reactors through long cruises and dangerous missions, made obvious candidates to be outstanding civil nuclear engineers. But as detailed above, the challenges of submarine PWR management differ in crucial ways to those that apply to a civil reactor, specifically in the emphasis placed on preventing the pressurizer from going solid. Control room operators who thought this way were intrinsically inclined to resist efforts to force water under high pressure into the reactor-pressurizer system, which was exactly what the ECCS was designed to do. This was why the operators in the Davis-Besse system had overridden the ECCS and shut down the HPIs when they should have left them open, just like the operators at TMI-2.

At around the same time that the Davis-Besse incident should have been flagging up warning signs for the nuclear power industry, an engineer named Carl Michelson uncovered a related flaw in the B&W 177FA PWR design. PWR operators were trained to use the presence of water in the pressurizer as a proxy indicator of water levels in the reactor itself, because in the archetypal PWR design, the pressurizer was on top of the reactor, and the two were connected. Thus, as long as there was any water in the pressurizer, the reactor must be full of water. This was useful because it meant that there was no need for expensive instruments to monitor the reactor water level. Michelson, however, had analyzed the design of the B&W 177FA reactor, and noted that it used a particularly tall pressurizer – too tall to fit directly on top of the reactor within the confines of the containment building. Instead the pressurizer was sited next to the reactor, with a sump-style inlet pipe connecting it to the reactor, and this meant that there would always be some water in the pressurizer, even if the reactor were empty of water. Thus, the operators of a 177FA reactor had no way to monitor the actual level of water in the reactor, and no way to know whether the blazing hot fuel rods might be in danger of getting left high and dry. Although Michelson's findings were passed on to both the NRC and B&W, as with the Davis-Besse incident findings, they got no further. As nuclear engineer James Mahaffey notes in his 2014 book *Atomic Accidents*, "There was a disaster, set up by a combination of policy and engineering, and it was waiting to happen."

In a later court case about liability for the accident, which pitched the operators against the reactor manufacturers, Zewe's testimony revealed how his training and experience had constrained his interpretations of, and response to, the control panel signals. Should he have realized, for instance, that the pressure continuing to drop, even as the water level in the pressurizer appeared to keep rising, indicated a problem with the PORV? "My training was that level and pressure would always trend together," Zewe testified. Asked whether this was always the case in his training on Babcock and Wilcox's simulator, he said, "Yes, sir, they always did." The trial featured a full-size mock-up of the control room panels, and Zewe used it to point out to the judge the PORV indicator lights. "The red light was out," he told the judge. "That indicated to me that the valve was shut. It's the only conclusion I could have reached." It had not occurred to him to question the accuracy of the instruments in the control room, he said. "I believed they were telling me what conditions were. I figured there must be a failure somewhere else in the system."

All of this intense drama, from the water hammer in the filter tank disabling the secondary coolant circulation, to the operators overriding the HPIs, was compressed into a few minutes. It was just after 04:00 on 28 March. In the control room, although they did not realize it, the operators were letting slip through their fingers the final chances to prevent catastrophic failure of the reactor. While they were focused on the apparent threat to the steam bubble in the pressurizer, they were not thinking about their true enemy: heat. Yet the reactor had been shut down just seconds after the incident began; why was it still pumping out a vast wattage of heat energy? The answer is decay heat. Each fission event produces a large amount of energy – 210 megaelectronvolts (MeV) – but only 89 per cent of this energy is immediately available. The remaining 23 MeV is effectively stored as potential energy in the fragments of fission and is released only as these fragments undergo radioactive decay. Decay is an exponentially reducing process, so that although it will take billions of years for the rate of energy release to reach effectively zero, the vast majority of the release is "front loaded". Decay heat tails off quickly, but immediately after fission in the core is shut down, decaying fission products are still producing significant amounts of heat. In a typical PWR reactor that produces 3,800 megawatts of heat when operating, decay heat means that, even after shutdown, the reactor is still producing 247 megawatts of heat, enough to melt solid rock – and certainly enough to melt fuel and cladding if they are not cooled. It takes an hour for the heat production to run down to about 57 megawatts, the level where fuel and cladding are safe, and so it is essential that a PWR has some form of cooling for at least an hour after shutdown. This is why Emergency Core Cooling Systems are so vital for PWRs. In the Davis-Besse incident, the reactor was running at only a fraction of its full power, and so the decay heat power after shutdown was relatively low and the fuel elements were not at risk. TMI-2 had been running at more or less full power, and immediately after the control rods went in it continued to generate at least 160 megawatts of heat. So long as the fuel rods in their zircaloy cladding were covered with liquid water, heat could be transferred away fast enough to stop them from melting. If they became uncovered, the fuel assemblies would quickly grow red hot, and soon heat up beyond the 1,650°C (3,000°F) threshold at which the zircaloy cladding starts to react with both the uranium inside the capsule and the superheated steam outside, and to break apart, with calamitous consequences. Uranium oxide and some fission products are water-soluble, so they can dissolve in the coolant and get out of the reactor vessel along with the escaping coolant.

With the PORV still open, primary coolant was still venting out of the reactor into the drain tank in the containment building. After 15 minutes it was full, and, as designed, it broached and overflowed, spilling water onto the floor of the containment building. This too was accounted for in the design, and the water flowed into a sump ditch at the lowest point of the building. When this filled up, pumps were supposed to transfer it to yet another large storage tank, still inside the containment building. Unfortunately, at TMI-2 the pumps had not been installed according to design, and they connected to tanks in the auxiliary building, outside the containment structure. If the sump pumps started to operate, there would be an extremely high risk of radioactively contaminated water escaping into unsealed areas, and thus of contamination escaping the plant.

By an hour after the incident began, 121,000 litres (32,000 US gallons) of coolant had vented through the PORV. With so much water missing from the primary system, steam voids began to build up, causing the primary coolant circulation pumps to cavitate (vibrate) so violently that the operators began shutting them down. By around 05:45 all four of the pumps had been shut off, and the reactor now had no supply of new coolant and no circulation of the remaining coolant. From around 06:00, as the remaining water boiled off and vented through the PORV, an ever-increasing portion of the fuel assembly rods was being left high and dry. Fire alarms started sounding, although there was clearly no fire. Zewe noticed gauges that showed air pressure in the containment building was climbing. Shortly afterwards, someone notified the control room that the sump ditch pumps in the containment area had switched on automatically (they ran for nearly 20 minutes before being shut down), and not long after this, a radiation counter in the corridor started to go haywire.

At 06:18, shift supervisor Brian Mehler, recently arrived in the control room, finally ordered the closure of the block valve on top of the pressurizer tank, sealing off the primary system. Not only was this a classic example of shutting the stable door after the horse has bolted, it was actually worse than that. With the block valve shut off, the last remaining avenue for the reactor to vent heat – by boiling off steam – was cut off, and the reactor began to melt down in earnest. Over the next eight minutes the top of the core collapsed. Radiation readings began to climb inside the reactor, and then alarms went off in the auxiliary building, where the mistakenly redirected primary coolant had ended up. Zewe now had no choice but to announce a formal Site Emergency. Over the next hour the station manager, Gary Miller, declared a General Emergency, and Zewe called the Pennsylvania Emergency Management Agency just after 07:00. At 07:20, three hours after the incident began, and believing that they needed to introduce cold water to condense steam bubbles, operators finally restarted the injection pumps. Unfortunately, it was now too late for this new coolant to arrest disaster; around half of the core had collapsed into a molten mass of fuel and cladding, around which formed a hardened crust. At the centre of this mass, where coolant could not penetrate, the conglomerate of melted material continued to swelter.

By 07:45 there were as many as 60 people in the control room. None of them was yet prepared to countenance the possibility that the core had been left uncovered by coolant, let alone that it may have melted down, and yet at that very minute the festering mass of molten debris in the centre of the reactor was undergoing a dramatic collapse. The crust around the base of it failed in one corner, and 20 tonnes of molten metal and fuel lava "relocated" – to use the euphemistic terminology of a report on TMI prepared by the Idaho National Engineering Laboratory – to the bottom of the reactor vessel, where it pooled against the steel bottom. Parts of this lava had reached temperatures of 2,760°C (5,000°F); under such intense conditions, even the 13 centimetre (5 inch) steel of the reactor vessel might have failed, and TMI might have been looking at an authentic China Syndrome. As a Smithsonian Museum account of the accident observes, "If this were known, or even merely surmised, drastic emergency measures, including evacuation of the region for miles around, would certainly [have been] ordered by the governor."

Just five minutes later, the governor received his first phone call about the incident. The possibility of a meltdown was not mentioned, but more alarmist voices would soon be raised. A traffic reporter for a Harrisburg radio station, WKBO, had overheard excited chatter on a police frequency and tipped off his news director, who in turn rang the main switchboard of the nuclear plant. The switchboard operator put him through direct to the TMI-2 control room, and as soon as the phone was picked up and the babble of panicked voices became audible, he knew there was a story breaking. The communications director at Met Ed quickly put out a bland statement simply acknowledging that the reactor had been shut down at 04:00 due to a pump malfunction, and that "[the unit] will be out of service for about a week while equipment is checked and repairs made".

Between around 07:45 and 09:00, the integrity of the reactor vessel, and thus the fate of much of Pennsylvania, was at greatest risk. The melting point of the steel walls of the vessel would have been about 1,500°C (2,732°F), while the lava-like melted core material might have been well over 2,000°C (3,632°F); a later metallurgical examination found that, if anyone had been present beneath the containment vessel, they would have seen the steel glowing red hot for at least an hour. Investigators later determined, however, that a mixture of zirconium and uranium oxides had formed a kind of ceramic layer that actually protected the wall of the reactor. The vessel held, and by around four hours after the incident began, the molten debris in the reactor began to cool. An estimated 18 billion curies of radioactivity – more than 100 times the amount released in the 1986 Chernobyl disaster – had been contained. In the control room, however, anxieties were only now beginning to really mount. At 09:00, radiation counters in the ceiling of the containment building were registering 6,000 rads per hour, an indication that dissolved fuel and fission products had escaped the reactor and were at large in the containment building. By 10:30, when NRC officials showed up in the control room, everyone was wearing respirators to guard against breathing in radioactive dust, as counters showed that contamination was leaking in.

At this point those in the operator room still believed that the reactor was full of water. They knew that forced cooling of the core, achieved by maintaining the circulation of primary coolant, had ground to a halt, but their understanding was that low pressure in the system had allowed steam bubbles to form, and that these were blocking the flow of water. Accordingly, they spent the morning trying to force water into the system to condense the supposed bubbles. In the afternoon, however, operators tried to decrease the pressure so they could use a low-pressure emergency cooling system, before reverting once more, by late afternoon, to injecting water under high pressure. Finally, at 19:50 in the evening, they were able to restart one of the reactor coolant pumps and restore forced cooling of the reactor.

Throughout Day 1 the utility company, Met Ed, was determined to minimize the severity of the accident. The governor's office pressed them for information so that they could make a statement to a nervous public, and in a morning press conference the Lieutenant Governor William W. Scranton told the people of central Pennsylvania that Met Ed, the plant's owner, had assured the state that "everything is under control" and that "all safety equipment functioned properly". As Governor Thornburgh reflected, "we later learned that it wasn't . . . and . . . that it didn't". He was scathing of Met Ed's response:

> The credibility of the utility, in particular, did not fare well . . . even when company technicians found that radiation levels in the area surrounding the island had climbed above normal, the company itself neglected to include that information in its statement to the public. The company had also vented radioactive steam into our air for two-and-a-half hours at mid-day, without informing the public . . .

In fact, there were several leaks/releases of gas with a low level of radioactive contamination during the first day. The first came early on, after some of the overflow coolant from the PORV drain tank was shifted by the sump pumps to the storage tank in the auxiliary building, which subsequently overflowed. Since this building was not sealed, small amounts of very mildly radioactive gas – mainly xenon-133 – may have leaked out of the vent stack in the roof. But the main source of venting was a process of "burping" that the operators undertook from early on Wednesday morning, in which they sought to relieve pressure from the makeup tank, a vessel that fed water into the primary coolant system. Bubbles of gas, such as radioactive noble gases, were threatening to disable this tank, and in order to release pressure, operators attempted to effect controlled releases from it to the waste gas decay tank, where it would be stored for long enough for the most active radioactive isotopes to decay and lose some of their potency. But the vent valve that connected these two tanks was known to leak, which meant that every burp would result in some gas escaping into the auxiliary building and up out of the vent stack, into the air. Periodically throughout Day 1, the plant operators undertook a laborious process in which someone in a full hazard suit went into the now radioactively contaminated auxiliary building, to turn on a compressor machine, which in turn would help the burped gas pass as quickly as possible from the makeup tank to the decay tank, and thus minimize the leakage from the vent valve. This worked in only short bursts, so the operator would have to repeat it several times at every visit. Periodic burping had been going on since 04:35, and between 11:00 and 13:30 – even as Scranton was talking – Met Ed engineers were at it again. According to local newspaper the *Patriot News*, there was yet another release at 17:00.

It is likely, thanks to a combination of factors including high-efficiency particulate air (HEPA) filters and charcoal filters on the vent stack in the auxiliary building, that almost all of the harmful radionuclides never left the auxiliary building. The most significant contaminant that did escape among the vented gas was xenon-133, a gaseous nuclide that would prove to be a major bone of contention in the days to come. Although it is radioactive and emits beta and gamma radiation, xenon-133 is relatively harmless to organisms because it is biologically inert, which is to say that it is highly unreactive and will not be taken up by the body or incorporated into any physiological processes. People might breathe it in, but they would most likely breathe it straight back out again without incident. And although concentrations at the point of emission might be high, it quickly disperses. Every five or so days, half of it will decay into non-radioactive caesium and harmlessly settle out of the atmosphere. None of these nuances, however, is likely to be apparent to the public in the wake of a major nuclear accident in their midst, and Governor Thornburgh had little patience for the mixed messages the utility was sending out. At 16:30 his lieutenant governor had to go back in front of the cameras and inform the public that "this situation is more complex than the company first led us to believe". He admitted that there had been, in the governor's words, "a release of radioactivity into the environment, that the company might make further discharges, that we were 'concerned' about all of this, but that off-site radioactivity levels had been decreasing during the afternoon and there was no evidence, as yet, that they ever had reached the danger point."

Unfortunately for all concerned, the NRC now began the first in a series of clumsy PR and communication missteps. Attempting to put out a statement minimizing any impact from the gas venting, they instead managed to make it sound like very slightly elevated radiation readings taken a mile away from the plant were due to "direct radiation coming from radioactive material within the reactor containment building, rather than from release of radioactive materials from the containment". To anyone with even a passing understanding of nuclear physics and nuclear plants, which included some of the reporters covering the story, it was clear that for radiation to pass directly from inside the containment building to a detector a mile away, it must pass through steel and concrete walls over a metre (3 feet) thick, and must be starting off at unbelievably high power levels. It would require something akin to an exploding nuclear bomb or a small star to project direct radiation that far. Accordingly, Walter Cronkite's CBS evening news bulletin led with the terrifying pronouncement that TMI had taken "the first step in a nuclear nightmare . . . radiation [is] so strong that after passing through a three-foot thick concrete wall, it can be measured a mile away". Fortunately, the general public in the Harrisburg area did not take Cronkite at his word, but worse missteps from the NRC were yet to come.

As Day 1 of the incident finally drew to a close, Governor Thornburgh was in for a sleepless night after remembering the book he had read about Fermi-1, *We Almost Lost Detroit*, and realizing that no one had yet briefed him on the question of whether or not a core meltdown had occurred. He would have been more troubled still to know what was happening within the hermetically sealed reactor vessel at that very moment, where zirconium atoms were stealing oxygen atoms from the steam in the void at the top of the core, leaving behind pure, highly explosive hydrogen gas, which was steadily accumulating in an ever-growing bubble at the head of the vessel.

Compared to the first day, the second day of the incident, Thursday, 29 March, seemed to be relatively uneventful. The NRC's "Report to the Commissioners and to the Public" (1980) on TMI, also known as the Rogovin report, after one of its authors, NRC director Mitchell Rogovin, would describe it as "The Interlude . . . a day for the drawing of deep breaths". In Washington, the NRC chairman Joseph Hendrie testified before a congressional committee. Asked by a congressman, "How close did we come to a meltdown, to a Chinese Syndrome?", Hendrie fired back, "Nowhere near". Meanwhile Met Ed told reporters that the plant was "stable" and that the controlled release of limited amounts of radioactivity into the atmosphere should soon be terminated. But official efforts to placate the public were hampered by the continuing ramifications of the NRC press release, with Tom Brokaw of NBC proclaiming that "the Nuclear Regulatory Commission in Washington says radiation penetrated through walls that were four feet thick and it spread as far as 10 to 16 miles from the plant". Governor Thornburgh recalled that "signs were popping up in grocery store windows proclaiming that 'we don't sell Pennsylvania milk'", in an echo of the Windscale accident (see page 85).

Thornburgh was aware that, despite public-facing displays of confidence, the scene at the plant itself was less reassuring. Operators were still struggling to bring the reactor to a cold shutdown, and in particular were battling with gas bubbles and high pressure in the primary system. "A certain air of apprehension was beginning to affect all those monitoring the process of recovery," Thornburgh later said. In his own daily press briefing he offered cautious reassurance and was startled when the NRC staffer who followed him promptly declared "the danger is over". Following coolant assays that revealed high levels of radiation, operators at the plant were beginning to realize that the damage to the core might have been much worse than they believed, and at the same time their anxieties were mounting about a suspicious bubble of gas in the reactor. The fact that, despite all their efforts injecting cold water, they had been unable to condense the bubble, indicated that the gas in the bubble was not steam. Also, at around this time, engineers realized that a pressure spike in the containment building on the Wednesday, which had barely registered amidst the general panic, had probably been due to a rapid burn of hydrogen that had escaped when the PORV was open. They were more and more certain that the gas bubble was hydrogen, and they could tell that it was growing; they would later learn it had now reached a volume of 570 cubic metres (20,000 cubic feet) – over half a million litres of gas.

Amongst the stress and fatigue of the control room, the threat of this hydrogen bubble loomed large. Hydrogen is highly explosive, and there was concern that the expanding bubble would either rupture the reactor vessel through sheer pressure, or explosively recombine with oxygen and blow the reactor to smithereens. In fact, both these fears were groundless. There was no free oxygen in the reactor vessel with which the hydrogen could combine, for precisely the same reason that the hydrogen was there in the first place: the zirconium had scavenged all the oxygen to create an oxide. Meanwhile the pressure tolerance of the vessel far exceeded the pressure that the bubble was exerting. But the operators were not thinking clearly, and fear was building. "Thus, Thursday ended on this somewhat edgy note," Governor Thornburgh remembered, "but it was a mere prelude to a Friday I will never forget."

Although the actual accident had happened more than 48 hours previously, Friday, 30 March was the day that would sear TMI into the consciousness of Pennsylvania and the world. Said the Rogovin report in resonant tones: "On Friday, Wednesday's apprehensions and drama will be reborn and multiply, and Thursday's calm will be shattered. Friday will bring back the accident like the back-half of a hurricane after the 'eye' has passed."

The initial drama was triggered by an infelicitous coincidence. Throughout Thursday, burping of steam from the makeup tank had continued. The fear of the operators was that, if the pressure in the tank got too high, a relief valve would trip, triggering a series of events that would impact on the supply of specially treated water they needed to replenish the primary coolant. At around 07:00 on Friday this scenario came to pass, and to limit the loss of coolant, operators in the control room agreed to perform a sustained venting of the makeup tank to bring down its pressure. They knew that this would entail further leaks of radioactivity into the air above the auxiliary building vent stack, but what they could not have known was that, about an hour into the venting, a helicopter would pass over the stack monitoring radiation readings. Passing directly through the plume of vented gas at the point of emission, it recorded disturbingly high readings of 1,200 millirems an hour. If this rate had been recorded anywhere else in the city, it would have been cause for alarm, but directly above the stack the plume was at its most concentrated, and would almost immediately be massively diluted as it dispersed.

What transpired next was what Governor Thornburgh called "a classic manifestation of . . . the 'garble gap' between Harrisburg and Washington", as the NRC's Washington-based Executive Management Team mistakenly thought that the readings had indeed come from off-site, and recommended an evacuation of all residents within an 8 kilometre (5 mile) radius of the plant. Compounding their error, they then passed on their recommendation through the wrong channels, with the result that word of it reached a local radio station in Pennsylvania, which was told that the governor might imminently be issuing an evacuation order for tens of thousands of people. The situation in Harrisburg was exacerbated still further by what Governor Thornburgh called "the mysterious tripping of an emergency siren that soon had hearts pounding and eyes widening all over the city". Already on edge, those who heard the siren reacted predictably: "People were throwing their belongings into trucks and cars, locking up their shops and homes and packing to get out of town," recalled Thornburgh. "If ever we were close to a general panic, this was the moment."

At 10:00, TMI-2 control room operator Ed Frederick had just come off shift after another long and weary night. He drove to Augie's Place in Middletown to pick up some sandwiches for his control room crew and was disturbed to see people piling into vehicles and driving away as though someone were after them. Inside Augie's, he found the proprietor shutting up shop; apparently there had just been an announcement on WHP Harrisburg radio to say that there had been a big release from TMI, and everyone within 5 miles needed to evacuate. Frederick rushed out to a pay phone and called the control room to find out for himself what was going on. Hearing his story, Zewe and another operator went straight to the nearest NRC officer, Inspector Jim Higgins, who was at that point in conference with Station Manager Gary Miller. "Frederick says they're evacuating Middletown and the surrounding area," Zewe told Miller, "What's wrong?" Miller was incredulous. "What are you people doing to us?" he barked at Higgins. The evacuation order would be swiftly countermanded by Governor Thornburgh, but, in the words of the Rogovin report, "not before letting loose fear that will roll around the area like a loose cannon, doing incalculable damage to the morale of this placid, stable region".

For the governor and his team, however, the evacuation issue was far from settled. "We began to wonder on our own," he later recalled, "if pregnant women and small children, those residents most vulnerable to the effects of radiation, yet relatively easy to move, should be encouraged to leave the area nearest the plant." They rang up NRC chairman Hendrie and asked him directly. His answer was not reassuring: "If my wife were pregnant and we had small children in the area, I would get them out, because we don't know what's going to happen." Thornburgh felt that he had to take a precautionary approach, and shortly after midday he issued a recommendation for pregnant women and preschoolers to evacuate an 8 kilometre (5 mile) radius of the plant, ordering all schools to close and evacuation centres outside the affected area to open. "Current readings," he told the public, "are no higher than they were yesterday [but] the continued presence of radioactivity in the area and the possibility of further emissions lead me to exercise the utmost of caution."

Oral testimony collected from locals by National Public Radio for their "I Remember TMI" project gives a flavour of reactions to the news. Local resident Vicki, who was seven months pregnant at the time and married to a local teacher, recalled:

> I will never forget the phone call I received from my husband. He emphatically told me NOT to go to my scheduled Obstetrics appointment at the Hershey Medical Center. He sounded scared and stated something was wrong. He told me many Hershey Medical Center physicians were picking up their children and taking them out of school. The physicians or their spouses did not say why and didn't take time to talk. They were in a hurry.

Vicki waited by the phone for her husband to call back, and eventually he contacted her to explain that he was being evacuated with those children from his class who had not yet been picked up. "Get a ride home to Annville and pack two suitcases," he told her. "We needed to be at least 60 miles away. If the reactor melted down, the ground would be contaminated . . . They kept saying everyone within a 10 mile radius from TMI needed to evacuate. We kept saying . . . what about 10.1 miles . . . so that's safe? Not an acceptable statement."

Middletown resident Joyce Corradi told the *Washington Post*, "I had the feeling that they didn't have a handle on what was going on." She and her family fled to her mother's home, 65 kilometres (40 miles) away, but when she got there, her 9-year-old son vomited what she described as "vile green slime". She recalled that while American doctors told her it was just down to stress or something he ate, a Japanese doctor supposedly told her it was "a classic symptom of radiation sickness".

The estimated total release of radionuclides at TMI was about 370 PBq (petabecquerels) For comparison, the estimated total releases from the Fukushima and Chernobyl disasters were 570 and 5,300 respectively, but at TMI the nuclides were almost exclusively short half-life, inert noble gases, mainly xenon-133. It is hard to see how anyone in the area could conceivably have received a dose of bioactive radionuclides capable of causing symptoms of acute radiation sickness, but many residents complained of wide-ranging problems including skin rashes, nausea and respiratory problems. Bill Whittock, a civil engineer who lived just across the river from the plant, was one of several locals who remember a "metallic taste" in their mouths immediately after the accident. Over 40 years later it is easy to see that such claims match the classic pattern of psychogenic symptoms triggered by stressful incidents, but at the time no one knew what to believe, especially when what should have been the most trusted authority, the NRC, continually undermined local efforts to prevent a mass panic.

That evening, loose talk from NRC officials in Washington continued to inflame public anxiety, with what Thornburgh called "an accurate but poorly handled statement" about the possibility of a meltdown. Officials in Pennsylvania shuddered when they heard Walter Cronkite leading the CBS Evening News by saying, "we are faced with the remote but very real possibility of a nuclear meltdown at the Three Mile Island atomic power plant." Again, Governor Thornburgh, this time joined by reinforcement in the shape of the NRC's director of nuclear reactor regulation, Harold Denton, had to give yet another press conference to try to calm the public. Behind the scenes, operators at the plant and officials in the NRC were indeed worried about the possibility of a meltdown, mainly because of the ongoing problem with the hydrogen bubble. Although scientists had now pointed out that there was no free oxygen in the vessel and hence no danger of an explosion, everyone was worried that the bubble might displace water from the reactor and thus expose the core (although, in reality, this ship had long since sailed). NRC chairman Hendrie, however, was developing a disturbing pet theory of his own. He was concerned that in the coming days free oxygen might evolve within the reactor system, and in particular that intense radiation in the system might cause radiolysis, radioactive splitting of water molecules, which would free oxygen molecules from the coolant water. This would bring an explosion into the realms of possibility. On Friday evening, he commissioned an independent study of the risks of such a development.

Also, concerns about the danger of a Windscale-type release of iodine-131 were still present, and it was late on Friday night when Food and Drug Administration officials started to ring around chemical manufacturers requesting

250,000 bottles of potassium iodide solution. Taking non-radioactive iodine just before exposure to iodine-131 can prevent the radioactive isotope being taken up the body and getting into the thyroid gland. In practice, while iodine-131 had been released inside the containment building, almost all of it immediately reacted with metallic structures in the building, bonding to them, and so it never escaped into the atmosphere.

Day 4 of the incident, Saturday, 31 March, brought yet another NRC-engendered scare story. Hendrie had been fretting about the hydrogen bubble all night, and at an afternoon press conference, although carefully worded, his anxiety about the issue was apparent. "With regard to the bubble in the vessel," he told the media, "that is a problem which is of concern and which we are working on very intensively at the moment . . . if enough oxygen over a longer period of time were evolved, why, it could become a flammable mixture."

As a brilliant nuclear physicist of great experience, Hendrie commanded a lot of respect within the NRC. "If Joe is worried," summarized one NRC official, "we had better take it seriously – not because we are scared of him, but because he is that good." But as the Rogovin report on TMI later reflected, Hendrie had got this one all wrong: "The hydrogen never explodes in the reactor vessel; it blows up, instead, in the media." Following up Hendrie's rather ominous briefing, an Associated Press (AP) reporter called various NRC sources to ask just how long it might be before "enough oxygen evolved". One source told him it might be just a few days; asked how many days, another source said two. At around 22:30, AP put out a report on the news wires: "Urgent . . . The NRC now says gas bubble atop the nuclear reactor at TMI shows signs of becoming explosive."

Once again, officials in Pennsylvania were speechless. The governor's press secretary, Paul Critchlow, stormed into the office of the nearest NRC official, Karl Abraham: "Karl, you should have told your people in Washington to keep their ----ing mouths shut, because the Governor is getting sick of it. You're causing a panic!" They were indeed. Many of the people in Harrisburg who heard about this story put together the words "hydrogen" and "nuclear" and understood that an H-bomb explosion was imminent. That night, 42,000 people left town, and by the next day, 135,000 people – about 20 per cent of the population within a 32 kilometre (20 mile) radius of the plant – had fled. In Goldsboro, near the plant, there were more reporters than residents; over 300 journalists were now on site.

From the Governor's point of view, the claim about a possible explosion was "totally groundless", and he and Denton scrambled to issue statements assuring the public they were safe. At 23:00 Thornburgh told the press: "there is no imminent catastrophic event foreseeable at the Three Mile Island facility, and I appeal . . . to all Pennsylvanians to display an appropriate degree of calm and resolve and patience". He also announced that, as a marker of confidence in the safety of the plant, President Carter himself would visit the plant the following day.

As for the hydrogen bubble, it was in fact shrinking as operators finally got the plant back under control. Meanwhile a B&W chemist, Don Nitti, who was running analyses on site, confirmed that the pressure exerted by the bubble would make it impossible for radiolysis to create any free oxygen. As far as the utility was concerned, the crisis was over.

On Sunday President Carter arrived. In addition to his taxing experiences during the NRX clean-up in 1952, Carter had trained as a nuclear engineer in the Navy, under Admiral Rickover. He was uniquely suited among politicians to understand the intricacies of the situation. As Governor Thornburgh later recalled,

> The President . . . and I toured the plant together – in full view of network television cameras. The image beamed around the world . . . had its desired effect. If it was safe enough at Three Mile Island for the Governor of Pennsylvania and the President of the United States, it had to be safe enough for the public.

Behind the scenes, however, Hendrie and some of the outside consultants he had tapped were far from sanguine about the safety of the plant. They were still profoundly concerned about the risk of a hydrogen explosion and were now armed with a controversial analysis that the reactor vessel would burst in that eventuality. Anxious voices in the NRC pushed Governor Thornburgh not to step down the civil emergency forces, but rather to prepare for a worst-case scenario and large-scale evacuation. They even advised him to have the National Guard on standby. But the governor was not having it. He had held a consistent line against spreading panic, and in what the Rogovin report described as "a fairly audacious gamble for a public official in office only a short time", Thornburgh ruled out what he called "a show of helmets".

Finally, on Sunday evening, Hendrie changed his mind about the risk of a hydrogen explosion. He had now heard from at least two sources that there was little risk that oxygen had built up, and in fact the latest readings from the plant showed that the bubble was shrinking fast. Hendrie and Denton were able to brief the governor that they could all stop worrying about the bubble, and two days later, on Tuesday, 3 April, Denton announced to the world, "the bubble has been eliminated, for all practical purposes". By 6 April, Governor Thornburgh was able to tell all those who had evacuated that they could "come home again".

In the aftermath of the disaster, the obvious concern was that radioactive contamination might increase the risk and incidence of cancer, and accordingly the Pennsylvania Department of Health maintained a registry for 18 years of more than 30,000 people who lived within 8 kilometres (5 miles) of Three Mile Island at the time of the accident.

When the registry revealed no evidence of unusual health trends, it was wound up. Over a dozen major, independent health studies, including a 13 year study on 32,000 people, have failed to find any evidence of abnormal cancer rates.

When the releases are put into context, it is easy to see why. The average radiation dose to people living within 16 kilometres (10 miles) of the plant was 0.08 millisieverts (mSv), roughly the equivalent of having a chest X-ray, and no single individual is believed to have received more than 1 mSv, a third of the annual dose a US citizen might receive from background radiation. It would take an acute dose of 100 mSv to increase lifetime cancer risk by just 0.4 per cent compared to all other causes. In 1996, when Harrisburg US District Court Judge Sylvia Rambo dismissed a class action lawsuit alleging that the accident caused health effects she concluded in her judgement:

> The parties to [this] action have had nearly two decades to muster evidence in support of their respective cases . . . The paucity of proof alleged in support of Plaintiffs' case is manifest. The court has searched the record for any and all evidence which construed in a light most favourable to Plaintiffs creates a genuine issue of material fact warranting submission of their claims to a jury. This effort has been in vain.

Yet despite this absence of evidence of any detectable public health effects, the long shadow of TMI still looms over any consideration of civil nuclear energy in the USA. The incident left a legacy of fear and suspicion out of all proportion to the actual harms that resulted. Part of the reason is obviously the chaotic and mismanaged public relations aspect of the incident, with confusing and contradictory reports and announcements, interspersed with panic-inducing official briefings that raised the spectres of evacuation, explosions and, of course, meltdown. The remarkable coincidence of the timing of the TMI incident to the release of the film *The China Syndrome* must have helped shaped public reaction. Twenty years on, Governor Thornburgh was still amazed at the coincidence, noting that "its script – incredibly – described a meltdown as having the potential to contaminate an area, and I quote, 'the size of the state of Pennsylvania'." An unnamed Met Ed official, quoted in a contemporary article in the *Patriot News*, griped, "I can't think of a worse time for this to happen – coincidental with *The China Syndrome*".

But was it simply a coincidence? As discussed above, the nuclear industry had already displayed its vulnerability to a number of relevant aspects: to accidents; to the tendency of plant operators to make them worse; to the failure to learn from them; and to a culture of secrecy and cover-up that made the public suspect the worst. A *New York Times* review of *The China Syndrome*, dated 18 March, 10 days before the TMI accident, canvassed the opinions of nuclear experts, including nuclear sceptic Daniel Ford. With remarkable prescience, Ford described the film as "a composite of real events [which] provides a scenario that is completely plausible". He went on to outline past failings that were all too shortly to be repeated:

> The failure of the pump support as occurred in the film could lead to a loss-of coolant accident that could set the stage for a meltdown. In fact, one utility had problems with a defective recorder, which led them to believe they had too much coolant rather than too little, and the Nuclear Regulatory Commission fined another utility a couple of years ago because one of its contractors falsified welding reports. All these things actually happened.

In other words, there was plenty of evidence to suggest that a TMI-style accident was in the offing, and perhaps every chance that such an accident would occur at a PWR somewhere in the USA, sooner or later.

As for TMI-2, the long battle to achieve cold shutdown was finally won at 14:03 on 7 April. No one yet had any concept of just how bad a mess had been made of the half-billion dollar reactor, nor of the reality that, rather than soon being back up and running, and earning money for Met Ed, it would cost over a billion dollars more to clean up. Quite the contrary; the experts were curiously reluctant to allow that a serious meltdown might have occurred. Although there was heated debate among them as to the condition of the core, a General Public Utilities engineer, who was in charge of much of the clean-up, later had to admit, "the most severe prediction was short of the mark". Most reactor engineers believed that, to the extent that there was core damage, it was to the cladding; the zircaloy tubes in which the uranium oxide fuel pellets were encased, and that damage to the pellets themselves was minimal. The Rogovin report, published in 1980, labours under the delusion that a meltdown had been averted when Mehler finally shut off the PORV at 06:18. Although it goes on to outline how a meltdown *might* have proceeded, allowing that "perhaps as much as half of all the fuel would have melted", it concludes that, while some fuel had melted and run down the channels between the fuel rods, "despite the amount of damage, a core meltdown, as normally considered, did not occur".

There was one crucial piece of evidence as to what had happened inside the reactor. About an hour and three quarters into the incident, when the crust of the initial molten agglomeration had ruptured, and molten uranium had spilled into the bottom of the reactor, neutron detectors beneath the vessel had lit up. The only possible source of such a high flux of neutrons was uranium fuel, and so engineers started to talk about an apparent "relocation" of fuel from the elements to the bottom of the vessel, but still no one was prepared to countenance the possibility of a meltdown.

Further progress towards understanding the extent of the damage would not be possible without actions that would have to be taken inside the reactor containment building, but the build-up of radioactive krypton gas in the building during the accident – gas that was still sealed in there – meant that it was not safe to enter. A year after the accident the authorities drew up a plan to vent the gas into the atmosphere. Like neon, krypton is a noble gas, which is chemically and biologically inert and thus, once dissipated in the atmosphere, not a threat to public health. Nonetheless the plan stoked a public furore. There was even the threat of unrest at the public hearings into the plan, and one of its opponents dressed up in a Superman suit and pretended to choke to death on the steps of the Pennsylvania capitol building. Governor Thornburgh got the Union of Concerned Scientists, described by him as "a well-known group of nuclear industry critics", to study the plan, and when they concluded that it posed no physical threat to public health and safety, public anxieties diminished. The venting went ahead, and 43,000 curies of radioactive krypton were released.

Assays of coolant from inside the reactor were inconclusive, and it took another two years for conditions in the containment building to improve enough for investigators to try more active interventions. First, investigators tried raising and lowering various control rods as means of gauging the situation within, but the results indicated little more than the possibility that the control rods themselves might be damaged. Finally, on 21 July 1982, a small video camera was lowered in through a channel cleared by sawing off parts of a control rod. One history of the TMI clean-up operation described how "there were many exclamations of surprise, disbelief, and confirmation", as the tiny camera was lowered into the almost impenetrably murky soup of old coolant that now filled the reactor. The first shock came when the camera was lowered to the level of the top of the core – or at least, where the top should have been. There was nothing there. After 1.5 metres (5 feet) of cable had been paid out, rubble came into view; the camera had reached the bottom of the cavity in the top of the reactor. A steel probe determined that there was a layer of loose rubble about 0.3 metres (1 foot) thick.

Further imaging of the core required novel solutions, and the Department of Energy's Idaho National Engineering Laboratory (INEL, although today known as INEEL, because the word "Environmental" has been added to the title) developed a sonar probe that provided ultrasonic mapping. According to the INEL, the sonar survey "established that the cavity in the TMI-2 reactor's core was substantially larger than had previously been supposed". Topographic mapping of the core, and preparation of 3-D models, helped illustrate just how extensive the damage was, to the top of the core at least. By April 1984 it was widely accepted that the upper 40 per cent of the core had been completely destroyed, but uncertainty remained about the rest. It would take years more to ascertain; in the meantime, after exhaustive tests, the crane above the reactor was declared operable and in July 1984, the top of the reactor vessel was lifted off so that removal of fuel and debris could begin.

Evidence from scans had suggested to many experts that the bottom of the reactor was filled with some sort of slag – the hardened remains of a once molten mass. In early 1989 it was finally confirmed that this slag was indeed mixed fuel and cladding material. Only now, almost a decade after the accident, did clarity emerge over the true narrative of what had transpired within the reactor, and in particular that the moment of peak danger had been when the initial mass in the centre of the core spilled 18,000 kilograms (20 tons) of molten material down the side of the vessel to pool at the bottom, threatening the integrity of the steel container. In order to obtain samples of this slag, the recovery team had to develop plasma arc cutting equipment that would work underwater and could be operated remotely from above the reactor. In early 1990 they carved out slices of the material and sent it for analysis, leading to the discovery, in 1994, that the molten fuel/cladding mix had congealed into a kind of ceramic, and demonstrating that the walls of the reactor vessel had withstood the intense assault, just as they were designed to do. What had become clear was that, in the TMI incident the reactor had come close to a China Syndrome yet, simultaneously, had not been in as much danger as had been feared. As a Smithsonian exhibit on the clean-up work summarizes, "In all, TMI showed that—contrary to common belief—a disaster inside a nuclear reactor does not necessarily lead to a disaster outside the reactor". Today the TMI incident is rated at level 5 on the INES scale, "Accident with Offsite Risks", thanks to the "on-site impact" being "severe core damage".

Clean-up workers had removed all the remaining fuel from the TMI-2 reactor in January 1990, but there still remained the problem of how to dispose of the nearly 8.5 million litres (2.24 million US gallons) of contaminated water generated in the course of the accident. The simplest solution was to let the water evaporate away, leaving behind the radioactive residue to be collected up and shipped away for long-term storage. This process began in early 1991 and was completed in August 1993. In September 1993 the body overseeing the clean-up, the Advisory Panel for Decontamination of TMI-2, held its last meeting. Stripped of over 99 per cent of its fuel, drained of its coolant, and with all the radioactive fuel and debris shipped to the INEL, TMI-2 was consigned to long-term, monitored storage until such time as TMI-1 would be decommissioned. In September 2019, with the economics of power production no longer favouring its continued operation, TMI-1 was finally shut down, and the whole TMI site has now begun its journey towards complete decommissioning.

But the clean-up of the reactor itself is only a small part of the story of the fallout from the TMI incident. For the manufacturer and operator of TMI-2 there was, inevitably, a legal wrangle, with Met Ed's parent corporation, General Public Utilities, suing Babcock & Wilcox for $4 billion.

The basis of their claim was that B&W failed to implement safety lessons learned from the Davis-Besse incident in 1977.

President Carter appointed a commission, chaired by John Kemeny, which attributed the blame for the accident to human error, stretching from the missteps of the Nuclear Regulatory Commission to the failures of the equipment manufacturers and the control room engineers. A particularly alarming finding was that similar failure of relief valves had happened several times, but that such incidents were covered up instead of being used to inform safety elsewhere. The shady culture of corporate cover-up and conspiracy peddled in *The China Syndrome* seemed to have been revealed in real life. The NRC today touts a list of improvements made and safety measures taken in the wake of TMI, from improvements in equipment and procedures, to better training and instrumentation. Crucially, they claim to have put in place institutions, practices and safeguards which better ensure that lessons learned from problems anywhere in the nuclear industry are applied across all plants. More broadly, TMI helped prompt a wider reassessment of the way in which safety systems should or can be designed in any industry. The incident prompted sociologist Charles B. Perrow to write his book *Normal Accidents*, which interrogates the very idea of fail-safe designs, pointing out that all complex systems have built-in levels of unpredictability. "No one dreamed",

Perrow wrote, "that when X failed, Y would also be out of order, and the two failures would interact so as to both start a fire and silence the fire alarm". In this sense, Perrow argues, something like TMI was both completely unpredictable and yet, at the same time, inevitable. The only solution is to develop ever more redundant and inherently safe backup measures and failsafe systems.

In the community around TMI, there have been conflicting emotions about the accident, the way it was handled, and the long-term effects on public health and confidence. Joyce Corradi, the Middletown resident who fled during the evacuation panic, and whose son was sick immediately afterwards, became an anti-plant activist. "There has been a great loss of innocence in this community as far as people in authority having the answers," she told the *Washington Post* in 1991, "I'm not sure people know what to believe". Another Middletown resident, Ann Trunk, who served as a member of the Kemeny Commission, agreed that "there is still an element of people who are frightened by nuclear power. I don't think people have changed their minds that much."

But there was also respect for the workers at TMI. A local man interviewed for the "I Remember TMI" oral history programme recalled, "A friend of mine worked at TMI. He never talked about what had happened, but, after the incident, he wore a T-shirt that read "TMI Safety Engineers: We Stayed on the Job and Saved Everybody's Ass".

And not everyone in town felt that there was a TMI "hangover". Ten years on, Middletown man Dennis Stover, who was in the local real estate business, described the original media coverage of the incident as "grossly distorted" and noted that there was plenty of evidence that the local community had moved on: "As quick as there are [homes] on the market, they're sold." Even Trunk admitted, "TMI is there, and so what? We've learned to live [with it]. Some people are comfortable."

Looking back, it is possible to see the story of TMI as one of systems put to the test. Some of the systems performed admirably – the mighty steel containment vessel, the filters that trapped the contamination; and some of them failed – the inherently non-redundant and non-failsafe, complex system of valves and pumps, and, most egregiously, the human systems of operator oversight and control, and public engagement and information. Ultimately those that have failed the most significantly have proved to be the system by which the public trusts the authorities to safeguard the operation of nuclear power plants, and, as a result, the entire system of civil nuclear power in the USA. Just as Windscale had forever contaminated the public perception of the nuclear industry in the UK, so the British newspaper the *Observer* described the effects on public opinion of the TMI incident as a "Nuclear Loss of Innocence". Over the five years following the accident, 51 orders for new nuclear reactors were cancelled. On 25 March 1979, the

USA had 140 operational nuclear reactors, with 92 under construction and 28 awaiting official approval. Although some plants that had already begun construction have been completed, and new reactors have since been installed at existing plants, not a single new nuclear power plant has been built in the United States since TMI.

Chapter 9

THE THIRD ANGEL: CHERNOBYL, 1986

"THEN THE third angel sounded his trumpet, and a great star burning like a torch fell from heaven and landed on a third of the rivers and on the springs of water." According to the biblical *Book of Revelation*, which vouchsafes this vision of doom, the name of this star was Wormwood, after a shrub that produces bitter toxins. "A third of the waters turned bitter like wormwood oil," the verse continues, "and many people died from the bitter waters." This grim prophecy would take on new significance in 1986, as news emerged of an appalling accident at the Chernobyl nuclear power station in Ukraine, then part of the Soviet Union. For the name of the power station is said to have derived from a local word for wormwood, and its atomic fire – related to the energy that fuels the stars – would indeed pour forth contamination to poison the air and water, claiming many lives. In fact, this derivation of the place name is a little tenuous, and Ukrainian wormwood is a different species to the biblical botanical, but there is no denying the apocalyptic nature of what happened at Chernobyl, or the bitterness of its environmental and human legacy.

The story of Chernobyl is a cautionary tale about the way that political and social systems can impact engineering systems, driving them to destructive failure and then skewing the response to that failure. Against this backdrop of dysfunctional authority and destructive forces, individual human stories of bravery and tragedy stand out. This chapter will seek to sketch the economic and political background to the disaster; explain its technical causes and immediate effects; detail the efforts to quench the blaze, evacuate local residents and entomb the reactor; examine the ongoing debate over the long-term health and environmental impacts; and touch on the historical consequences of Chernobyl.

In 1986, the Chernobyl nuclear power plant, sited in Ukraine not far from the border with Belarus, was one of the crown jewels of Soviet industry. The third most powerful nuclear power station in the world, it seemed to be a rare bright spot amidst an increasingly gloomy outlook for a stagnating economy, and accordingly the plant and its ilk would assume central importance at the showpiece political event of the year, the 27th Communist Party Congress, held in Moscow from 25 February. The Congress provided a chance for the dynamic new General Secretary of the Communist Party, and thus the new leader of the Soviet

Union, Mikhail Gorbachev, to lay out his plans for coping with the crushing economic crises blighting the USSR. By the Party's own estimates, annual GDP growth had plummeted from 10 per cent in the 1950s to just 4 per cent; a more realistic assessment by the CIA suggested it was probably as low as 1 per cent, while the Soviet military was mired in an intractable and ruinously expensive conflict in Afghanistan. The USSR was struggling to fill supermarket shelves with basic foodstuffs, let alone keep pace with rapidly increasing American defence spending. Gorbachev's plan to overcome this pervasive *zastoi*, or stagnation, was a new policy of *uskorenie*, acceleration: a drive to achieve greater productivity by leveraging the power of science and technology. Central to this policy, with its wildly ambitious target to double Soviet GDP over the next 15 years, would be atomic power generation, with fossil fuel plants replaced by ever more powerful nuclear power stations. The Ministry of Energy and Electrification had been set a target of a two-and-a-half-fold increase in the production of electricity by nuclear reactors over the next five years. This was plainly far-fetched, given that in theory it took at least seven years for a reactor to move from initial design to completion, and much longer in practice. But this was how the Soviet system worked; the minister was happy to tell his superiors what they wanted to hear, and then pass these undeliverable demands on down the chain of responsibility.

One of those at the sharp end of this daisy chain of unmeetable expectations was the director of the Chernobyl plant, Viktor Briukhanov. He had overseen the construction of four powerful reactors, the plant around them, and the brand-new city which housed the workforce that constructed and ran the plant. Two more reactors were under construction, but Gorbachev's address, which Briukhanov had witnessed in person as a delegate at the Congress, meant that this would have to be just the beginning of an even greater push to expand the capacity and production of his already sprawling plant. Speaking to a Ukrainian reporter on 6 March, the last night of the Congress, Briukhanov was moved to comment, "We must hope that [this drive to boost nuclear power] will also promote greater attention to the reliability and safety of atomic energy generation at our Chernobyl station in particular. That is most urgent for us." This remark was cut from the published interview; Soviet authorities did not care for criticism, no matter how carefully couched, nor for anything that smacked of negativity – *especially* if it was actually simply realism.

The station to which Briukhanov now returned lay in a forested region, about 130 kilometres (80 miles) north of Kiev and 20 kilometres (12 miles) south of the border with Belarus. The city of Chernobyl (actually Chornobyl in Ukrainian – Chernobyl is the Russian version of the place name) is an ancient port on the Prypiat River, with a population then of only around 14,000. It lay about 12 kilometres (7 miles) to the south-east of the power plant, while the new city of Prypiat, home to around 50,000 construction and plant workers, and their families, lay just 3.5 kilometres (2 miles) to the north of the plant. Each of the four reactors already in operation had the capacity to produce up to 1,000 megawatts (MW) of electricity, by producing up to 3,200 MW of thermal energy, i.e. heat. To distinguish between these two ways of measuring the reactor's power output, the notation *e* or *t* is used. Each of the Chernobyl reactors could thus be described as having a power level of up to 3,200 MWt.

The reactors were all RBMK-1000s, the workhorse of the Soviet civil nuclear reactor fleet. RBMK stands for *reaktor bolshoy moshchnosty kanalny*, or high-power channel reactor. The design was allegedly first inspired by a Soviet comedian's skit about not letting the energy of a twirling ballerina go to waste, in which the performer is hooked up to a dynamo to produce electricity. In the case of an RBMK, the "ballerina" is a huge stack of graphite blocks through which pass channels for fuel rods and pipes carrying water. The graphite moderates neutrons emitted by uranium nuclei in the fuel, slowing them down enough to be captured by other nuclei, and thus sustaining a nuclear reaction. The water cools the core, absorbing heat and carrying it to steam generators, which in turn produce steam to drive turbines. Note that the water does not come into direct contact with fuel, and runs to the steam generators in a single loop, unlike PWRs, which have a primary and secondary cooling loop. Such reactors are also sometimes known as light water graphite reactors (LWGRs).

The most prominent risk built into such a design is known as the positive void coefficient, which refers to the susceptibility of the reactor to a damaging feedback loop. Water reactors, like the boiling water reactor (BWR) at SL-1, or the PWRs of Three Mile Island and most other US nuclear power stations, use water as both moderator and coolant, with a concomitant self-limiting effect on power output, because as power levels rise and the water heats up, it expands and moderates less efficiently, decreasing reactivity and lessening power (see page 221). The effect on the reactivity of the reactor is described by its void coefficient. BWRs and PWRs have negative void coefficients. An RBMK reactor, however, does not have this in-built, self-limiting safeguard; when the water in the coolant heats up and forms steam voids, the effect is to increase the reactivity of the core, with the potential to cause a dangerous positive feedback loop. Its void coefficient is positive.

Soviet decision makers, however, were not interested in void coefficients. Far more eloquent, to their ears, were the RBMK's advantages. Compared to its main competitor, the VVER (the Soviet version of the PWR), RBMKs were twice as powerful, and cheaper and easier to build and operate. They could use minimally enriched uranium (just 2–3 per cent) as fuel; they could produce plutonium isotopes for weapons; it was possible to swap out fuel rods while the reactor was running at full power, reducing downtime; and they could be built at scale to produce massive power outputs. By 1982 more than half the electric power in the Soviet nuclear power industry was produced by RBMK reactors.

Apart from the positive void coefficient, the RBMK came with other major safety risks, since it contrived to combine some of the worst characteristics of other reactor designs. The graphite core posed a major fire risk, while the water coolant system introduced the threat of steam explosions. Worst of all, RBMKs were built without concrete containment structures. The ultimate reason for this was probably the sheer size of the reactor units (the graphite cores alone were 14 metres/46 feet in diameter and 7 metres/23 feet high), along with the height of the fuel element handling rigs that had to sit atop the reactor, which needed 35 metres (115 feet) of clearance, and which in turn meant that any containment structure would have to have been impossibly vast. But the Soviet nuclear industry claimed that the expensive structures were not necessary because the design was so safe. Gorbachev himself would later recall being told "that an RBMK reactor could be installed even on Red Square, since it was no more dangerous than a samovar [tea urn]".

Not everyone in the Soviet nuclear establishment agreed with this assessment. A significant dissenting voice was the chief designer of the RBMK, Nikolai Dollezhal, who pointed out that while the Americans had, by the early 1970s, increased spending on their reactors by a factor of eight, in order to institute exhaustive safety measures, no such increase in spending had happened in the USSR. He argued that RBMKs should be built only in remote eastern areas, and not in more densely populated European parts of the Soviet Union.

In 1975, this more cautious assessment was graphically underlined by a serious accident with an RBMK reactor near Leningrad, one which would prove to have massive implications for Chernobyl, and which should have led to lessons being learned and design flaws being rectified. During a routine low-power operation, voids formed in the reactor's coolant, leading to runaway reactivity, which could not be controlled through the usual control rod procedures. When the reactor was scrammed for emergency shutdown, there was a dangerous spike in reactivity before the reaction was killed, and it turned out that it was due to the design of the scram rods. Although most of the length of the rods – the active region – consisted of neutron-absorbing boron, the tops and bottoms of each rod were hollow tubes, while the tips of the rods were capped with non-neutron absorbent graphite, there to act as a lubricant. As the rods were inserted, the tips were the first parts to enter the core, and for the period when they were the only material present in the control channels, these tips had an effect opposite to that intended, increasing the reactivity of the core rather than reducing it. Thus, as insertion began, there was a momentary spike in reactivity: a positive void effect. Whereas scram rods in a Western reactor design typically take only around 3 seconds to fully insert, the scram process in an RBMK was drawn out, taking up to 20 seconds, of which up to

5 seconds would cause a positive void effect. Adding an extended positive void effect to a reactor already on the edge of criticality could be disastrous.

This was exactly what happened at Leningrad, where scramming the reactor caused a positive void effect, with a concomitant spike in reactivity, which resulted in the meltdown of the fuel in one of the channels, releasing uranium and other fission products into the core. This necessitated a costly "cleaning" of the reactor, in which the channel was flushed out with nitrogen gas and the waste was simply vented through an exhaust pipe, releasing 1.5 million curies of radioactive contamination into the environment, just 50 kilometres (30 miles) from the major city of Leningrad (now St Petersburg). The International Atomic Energy Agency's (IEAE) "safe" level of contamination is set at 5 curies per square kilometre. No one knows what impact this massive, uncontrolled release had on the people or ecology of the area, but 1.5 million curies are enough, for instance, to contaminate almost 10 billion litres (2.6 billion US gallons) of milk.

Under an open and functional regulatory regime, the Leningrad incident would, at the very least, have caused an instant shut down of all RBMK reactors until the design of the control rods could be overhauled, and all operators would have been diligently drilled in the lessons learned from the accident. In the Soviet Union, the incident was hushed up to the extent that even some operators present in the Leningrad control room when it happened were never told what the cause was, and news of it was suppressed. Although other plants were advised to modify their control rods, they were not told why, nor how urgent it was, and accordingly at Chernobyl there was no great urgency over the issue; it was added to a list of recommended tasks to be performed during the increasingly infrequent maintenance breaks. The rod redesign had taken place at Unit 3, for instance, but not at Unit 4.

At the time of the Leningrad incident, Chernobyl was still under construction. When Briukhanov had arrived at the small town in 1970, there was little to see to the north but forest and low sandy hills. By September 1977 when Unit 1 was connected to the grid, he was able to declare, "[this] year will go down in the history of Soviet atomic energy as the year of the birth of an energy giant on the Prypiat". Under his stewardship, Unit 2 was connected to the grid in December 1978, Unit 3 in December 1981, and Unit 4 in December 1983 (Units 5 and 6 were still under construction). These December dates were significant; they reflected the immense pressure from the authorities to meet annual targets. Accordingly, every sinew was strained to ensure that a project could be declared complete before the year-end cut-off. In the rush to finish, even more corners might be cut than usual, particularly concerning given the generally poor level of workmanship that had gone into the plant.

KGB surveillance of the process had uncovered numerous instances of sub-standard work, such as walls 15 centimetres

(6 inches) off true, and concrete made with the wrong aggregate, which subsequently had large holes in it, while there were similar problems with the quality of parts and material supplied. It was ultimately Briukhanov's responsibility to ensure construction standards were met, but it was also deemed his fault if the plant did not meet deadlines. This double-bind reflected a fundamental flaw in the entire, command-economy system: the government was both contractor and client. Briukhanov and everyone else in authority needed to pretend that work was being done on time and to satisfaction, when really it was late, shoddy and dangerous, but they also needed to ensure that the plant operated successfully, which required decent construction.

So far Briukhanov had won plaudits and prestige for seemingly pulling off this balancing act, not just at the plant, but also in the parallel task of overseeing the construction of the rapidly growing city of Prypiat. With its civic slogan "Let the Atom be a Worker, Not a Soldier", Prypiat revolved entirely around the power plant, although it would soon discover that which gives life can snatch it away in an instant. Planners had originally designed it for 12,000 inhabitants; by 1986 it was home to around 50,000, mainly young people, often starting families: 1,000 newborns arrived every year. But the growing city siphoned money away from the plant, causing Briukhanov to grouse to a reporter in early 1986, "We get used to the abnormal and begin to accept it as something like the norm. That is what's so terrible!" He seemed all too aware of the risk this might pose to the operation of the plant: "In those second-rate circumstances, the main thing is to ensure the reliability and safety of our work . . . we are no ordinary enterprise. God forbid that we suffer any serious mishap – I'm afraid that not only the Ukraine but the Union as a whole would not be able to deal with such a disaster."

The heart of the power plant was a contiguous complex of vast buildings. A long turbine hall, 32 metres (105 feet) high, connected all four units, with Units 1 and 2 at one end, standing slightly apart from one another, and Units 3 and 4 housed in a huge building, 72 metres (236 feet) high, which extended out on either side of a shared exhaust chimney. This stack, with its red and white stripes and cat's cradle of supporting scaffolding, would become the icon of the Chernobyl disaster. Next to the plant was a huge cooling pond, created by diverting part of the Prypiat River.

For several years Chernobyl had exceeded its electricity production quotas, and despite the suspect workmanship and materials, the plant had mostly avoided incident. There had been an accident with a burst fuel channel in Unit 1 in 1982, but Soviet censorship rules made it illegal to report in the media. What was widely reported, however, were comments by Chernobyl chief engineer Nikolai Fomin, in a Soviet English-language publication, that the plant was so safe and clean that the cooling pond at the station was used for breeding fish – and it was true that while it was officially forbidden to fish there, many plant workers did.

The colossal reactor hall of Unit 4 consisted of little more than a thin roof over a huge space. Inside it, sitting in a reinforced concrete pit, was the reactor with a cylindrical stack of graphite blocks surrounded by a cylindrical water jacket, and capped at top and bottom by steel and concrete biological shields. The upper biological shield was a massive disc-shaped sandwich of sand and concrete between 4 centimetre/1½ inch-thick steel plates, 3 metres (10 feet) thick and 17 metres (56 feet) across. Figures for the mass of this shield vary between sources. According to the World Nuclear Association, it weighed 1,000 tonnes. It was nicknamed the *Pyatachok*, or "5-kopek piece", a reference to an oversize coin from Tsarist times, allegedly designed to make it easier to handle while wearing gloves in the bitter cold of a Russian winter. Running through the shield were channels for fuel and control rods and coolant lines. After the accident, the shield would be designated Element "E", nicknamed Elena, and the bundles of rods and tubes still attached to it would become known as "Elena's hair".

After two and a half years of operation, the reactor in Unit 4 had nearly reached the end of a fuel cycle, the period over which a load of uranium fuel was used up to the point where it needed to be replaced. In practice this did not mean that more than a fraction of the approximately 200 tonnes of uranium in the core had actually undergone fission, but it did mean that within the fuel elements, there had built up a considerable amount of unwanted fission products, and it would soon be time to swap them out for fresh fuel. In addition, Unit 4 was in need of maintenance. The previous year, the power plant had exceeded production quotas by reducing the number of shutdowns for repairs, and now there was a backlog. Accordingly, Unit 4 was scheduled for shutdown on the weekend of 26/27 April.

This was the weekend before the much-anticipated May Day holiday, in preparation for which a new funfair had been constructed in Prypiat. Constructors had recently completed the installation of bumper cars and a yellow Ferris wheel, and the fair was scheduled to open in a few days. Workers who originally came from the surrounding villages were preparing to return to their relatives' homes to spend the weekend helping with agricultural duties such as planting potatoes.

As part of the shutdown procedure, and in order to take advantage of low power levels, the reactor would often be put through tests. On this weekend it was planned to run a test of a new idea to improve safety by boosting the emergency cooling response to an abrupt scram or other shutdown. Since such a shutdown also cut off the supply of electricity that usually powered the coolant pumping system, and there was a gap of approximately 45 seconds before the back-up diesel generators kicked in to provide reserve power, engineers wanted to try out a scheme to fill this gap. The concept was to utilize the inertia of the steam turbine system to continue to generate enough electricity to plug the gap; after all, while fission itself might come to a dead halt in the event of a scram, the rest of the system did not. The massive dynamos in the turbine hall would keep spinning, albeit at a rapidly reducing rate, for some time, generating some degree of electricity. Just how long and how much were unknown factors that engineers wanted to explore in the test.

A programme was drawn up for the test, and it included a schedule for reducing the power levels and taking advantage of what was supposed to be a brief window between reaching low power and total shutdown. To achieve low power, operators had to throw in all but 15 of the control rods, a major risk to reactor stability, according to reactor's original designers. In the words of Serhii Plokhy, author of the landmark book on the tragedy, *Chernobyl: History of a Tragedy* (2018): "With reactors, as with airplanes, takeoff and landing are the most challenging moments". Indeed, according to the manufacturer's instructions, any reactor that proceeded this far should be shut down immediately. But the test programme took priority over the poorly observed manufacturer's strictures.

The original programme called for the entire shutdown test to be finished by 10:00 on Friday, 25 April, which meant starting the evening before. In the event, the power reduction process was not begun until the early hours of 25 April, dropping to a level of 1,600 MWt by 04:48. At this stage only 15 control rods were not engaged, but no one considered following the manufacturer's instructions and shutting down completely. Instead, the next step in the programme called for the emergency cooling water supply to the reactor to be disabled, a laborious process involving manual closure of a series of valves. It raised obvious risks, but operators assumed that there was a vanishingly small chance of the main water supply being interrupted during the test. By 14:00 the emergency water system was shut down. In less than 20 minutes, operators were supposed to commence reducing the reactor power level still further, reaching shut down via a level of just 700 MWt.

But at this crucial juncture came a call from the dispatcher at the Kiev area power grid HQ. This was the office responsible for controlling the supply of power to the grid in Kiev and the surrounding region, and their orders took precedence over everyone else's, except in the case of an extreme emergency. An outage at another power station meant that the power grid controllers had determined that they could not afford for Unit 4 to go offline. The test would have to wait until that evening, and the reactor would have to remain in its unsatisfactory state – just short of shutdown with the emergency cooling system turned off – until then. Among other things, this meant that the shift which would be on duty in the control room for the actual test was not the one that was familiar with the test programme.

The evening shift supervisor, Yurii Trehub, was not happy about this situation, later complaining bitterly, "I'm surprised that there could possibly be such a turn of events – a dispatcher taking command of an atomic power station." His superior told him that the grid dispatcher was expected to give permission for the test to continue at 18:00, and so they decided not to cancel the test or turn the emergency water supply back on. Instead they waited; and waited, for no word came from the grid dispatcher. At 20:00 Trehub called a superior, who told him to speak to the plant's deputy chief engineer, Anatolii Diatlov. But at first Diatlov could not be located, and when he did get through, Trehub was told to wait for Diatlov to arrive in the control room before proceeding with anything. Finally, at 21:00, the dispatcher gave permission for the shutdown to proceed at 22:00. Now Trehub had only to wait for Diatlov to arrive, but Diatlov knew that the electrical engineers, whose idea the test had been, would not turn up until 23:00, and so that is when he got there. At midnight a new shift took over in the control room, and Diatlov was finally ready to begin.

The man in charge of the new shift was a younger, much less experienced engineer, Aleksandr Akimov. Allowed no time to familiarize himself with the test programme, and with Diatlov breathing down his neck, Akimov struggled to orchestrate the complex operations now required. With over 500 dials and controls, finessing the reactor, especially at low power, was akin to playing an organ or conducting an orchestra. Over the next hour and 20 minutes, with the help of and sometimes under instruction from Trehub, who had hung around to observe, Akimov juggled control rods, water pumps and cooling systems as the team struggled to control fluctuating power levels.

At times the power level of the reactor dipped almost to zero, as the operators tried to keep it steady at 200 MWt – much lower than the 730 MWt proscribed in the test schedule. At such low power levels, margins are much finer and every decision can have unintended knock-on consequences. For instance, turning on reserve pumps to increase the flow of cooling water, as called for in the test sequence, changed the ratio of steam to water in the system. Water absorbs some neutrons, while steam does not, so the extra cooling also had the effect of reducing reactivity. As a result, the operators had to shut down the reserve pumps. Even worse was the effect that low power running had on the build-up in the fuel rods of xenon-135, a fission product which acts as a "neutron poison" – a substance that soaks up neutrons and undercuts the reactivity of the core. During normal running, the neutron flux of the reactor burns off xenon-135 at the same rate that it is formed, but when the reactor is reduced to low power, it is stuck with the xenon load from earlier high power running, but without the neutron flux to burn it off. In such circumstances the xenon-135 can quickly stifle – or poison – fission, killing the reactor stone dead. In such a scenario, it will take 45 hours for enough of the xenon to have decayed away to allow the reactor to be restarted. In order to keep the reactor from flatlining, Akimov and Trehub had to pull almost every control rod out of the core, until just 6 out of 211 were inserted. This was profoundly dangerous, for it seriously limited the operators' room for manoeuvre, and greatly worsened the instability of the reactor.

At every turn, every indicator should have directed the operators to shut down the reactor, but Diatlov was single-mindedly focused on completing the test programme. With the imperative to complete the programme overriding all other concerns, Diatlov and his subordinates agreed between them to proceed with the test, despite the low power and other issues. Even as Akimov prepared to give the order to start the test, data from the computerized monitoring system revealed an alarming spike in reactor power levels; with the reserve pumps now turned off, water in the cooling system was boiling, forming steam voids. The abrupt phase change from water to steam also meant an abrupt cessation of neutron capture and a corresponding spike in the reactivity of the core. One of his deputies, Leonid Toptunov, called out the data but Akimov was too intent on the imminent test. At 01:23:04, the turbine test began, and the turbine system was disconnected from the reactor system. The turbogenerator coasted down, exactly as expected, and after 36 seconds Akimov gave the command to shut down the reactor. The test was over, but the nightmare was just beginning.

During those 36 seconds, with the turbine not drawing power, the steam voids expanded rapidly and the reactivity spiked: the core had begun to run out of control. Akimov eyed the AZ-5, the scram or emergency shutdown trigger. "AZ" comes from the Russian acronym for "Rapid Emergency Defence". Toptunov called out that the power level was rising too fast. Akimov pointed his finger, and called out, "Press the button".

The AZ-5 button activated the scram system: 178 control rods, each 7 metres (23 feet) long, were driven into the core at a speed of 40 centimetres (15 inches) per second. At the same time, the 205 withdrawn control rods were thrown in as well. As the graphite tips of all the rods entered the top of the core, they combined to generate a massive positive void effect. The reactivity of the core leapt, generating a blast of intense heat that fractured the cladding of the fuel rods, blocking the channels and jamming the control rods when just a third of their lengths had been inserted. Over the next seven seconds a tsunami of fission events overwhelmed the stricken reactor, and the power level rocketed, from 200 to 500 MWt, and then to 30,000 MWt – 10 times its operating norm.

The huge blast of excess neutrons burned through the xenon-135 poison. As it burned off, the full power of the reactor was unleashed: the xenon had acted like the handbrake on a revved-up car, the wheels of which are spinning furiously in place, so that when the handbrake is released the car rockets forward. The flood of thermal energy disintegrated the zircaloy cladding of the fuel rods and molten fuel tablets came into direct contact with the coolant water, which flashed into steam. At 01:23:44, Unit 4 of the Chernobyl power station suffered a colossal steam explosion.

In the control room, the operators had allowed themselves a sigh of relief when the AZ-5 was pushed, believing that the shutdown would now proceed and the drama was over. Instead they were shocked by an immense noise. "That roar was of a completely unfamiliar kind," operator Razim Davletbaev later recalled, "very low in tone, like a human moan." After the roar came a series of shocks, of increasing strength, followed by the sound of the blast. A few seconds later came a second, more powerful blast. "The floor and walls shook violently," remembered Davletbaev, "dust and bits of debris fell from the ceiling, the luminescent lighting went off, semi-darkness descended, and only emergency lighting was on." "Everyone was in shock," said Trehub. "Everyone stood around with long faces. I was very frightened. Complete shock." Had there been an earthquake? What could have happened?

What had happened was a catastrophe. The first explosion had blown the enormous Elena shield through the roof of the reactor hall, ripping out, as it went, the guts of the reactor, including the tops of fuel channels and many of the coolant circulation pipes. It came back down on top of the reactor, but tipped over on its side, leaving a 5 metre/16 foot-wide gap that exposed the core, through which radiation could now spew forth. With the coolant supply abruptly cut off, the remaining fuel was left to overheat even more, with the white-hot zirconium cladding scavenging oxygen from the water to create hydrogen, even as the water tanks beneath the reactor vaporized into a huge blast of steam. The combined mixture of superheated steam and hydrogen exploded, blowing apart much of the Unit 4 building and a large chunk of the graphite core, propelling chunks of graphite and fuel up into the air to land on and around neighbouring buildings, and setting fire to the remaining graphite, so that radioactive smoke and fumes would cascade out into the atmosphere. With no containment structure in place, to guard against just such a calamity, there was nothing to stop the escape of radioactive nuclides into the environment. The absence of this crucial feature completely changed the course of the disaster. According to Bernard Cohen, professor of physics and radiation health at the University of Pittsburgh, "Post-accident analyses indicate that if there had been a U.S.-style containment, none of the radioactivity would have escaped, and there would have been no injuries or deaths."

Over a quarter of the 1,200 tonnes of graphite in the reactor, along with up to 50 tonnes of uranium, were ejected or vaporized in the second explosion, along with 360 kilograms (794 pounds) of fission products, including extremely dangerous strontium and caesium. They were blasted up to a kilometre into the atmosphere. The remains of the reactor inside the vault resembled a volcanic crater, a pit carved out of graphite, most of which would burn up over the next 10 days. Later that night, a shocked Diatlov would explain to the night shift manager, "we pressed the AZ-5 button, and 12 to 15 seconds later the unit exploded".

One of the few people actually to see the explosion was a 16-year-old girl, Natasha Timofeyeva, on her way home from a party in the village of Chamkov, just across the border in Belarus. She saw a bright flash from beyond the giant chimney. Nearer at hand were fishermen around the cooling pond, two of whom were sat just 260 metres (853 feet) from Unit 4. Perhaps used to strange noises and disturbances from the plant, they kept on fishing.

Not far from the plant was the station house of Specialized Military Fire Department No. 2. The firemen on duty heard the two explosions and saw a fireball and a mushroom cloud of smoke rising above Unit 4. They leapt into their fire trucks and just five minutes after the explosion they had arrived at the plant, where they were shocked to see the extent of the damage to the roof and walls of Unit 4. They could see flames. Firefighter Leonid Shavrei later recalled, "My hair stood on end." His commander, Volodymyr Pravyk, got on the radio to his dispatcher: "Call everybody, everybody!" An alert of the highest level was circulated, mobilizing all firefighting units in the Kiev region.

From their position outside the reactor buildings, the firefighters were uncertain as to the actual site of fire; it looked as though it could be the roof of the turbine hall in flames. When they climbed up there, they discovered that, contrary to regulations, the roof was covered with flammable bitumen. They also found, Shavrei reported, that "the whole roof was littered with luminous, silvery pieces of debris of some kind. We kicked them aside." Incredibly, the firefighters with responsibility for the Chernobyl power plant had received no training about how to deal with radiation, and no clue that they were kicking pieces of graphite and nuclear fuel that had come from inside an active reactor, let alone that each one was blasting out lethal doses of radiation. All they knew was that the blazing hot chunks of material kept starting fires.

A second crew that arrived from Prypiat shortly afterwards went to tackle a blaze on the roof of Unit 3. Pravyk joined them, and they battled to stop flames spreading to the exhaust stack. When they climbed back down less than half an hour later, they met Pravyk's commanding officer, Major Leonid Teliatnikov, the head of Fire Department No. 2. "There were seven men with [Pravyk], all in bad shape and feeling sick," Teliatnikov recalled. They were suffering the immediate symptoms of acute radiation sickness. He ordered them into an ambulance. Pravyk must by now have gleaned an inkling of what was afoot, for he asked someone to call his wife in Prypiat and tell her to shut her windows. Another crew commander, Lieutenant Viktor Kibenok, told a newly arrived doctor, "Something is making my boys feel a bit sick." The doctor, Valentyn Belokon, was soon confronted by a stream of patients reporting with similar symptoms: headache, nausea, dry throat, vomiting. One young workman was moaning, "Horrors! Horror!" Initially Belokon suspected alcohol poisoning, but after sending a stream of sufferers to the hospital, he realized what must be going on, and called his superiors to ask for potassium iodide, which might at least guard against thyroid contamination.

Struggling to get water for the hoses, firefighter Petr Shavrei (Leonid's younger brother) led his truck to the nearby cooling pond, but the route was blocked with debris, including rods that punctured the truck tyres. "I took metal rods out of the wheels with my hands and kicked them out with my feet," reported Petr. "Then the skin peeled off my hands. . . ." Soon Petr, Leonid and a third firefighting Shavrei brother, Ivan, were all stricken with radiation sickness. "I was retching and felt terribly weak," Petr said later, "My legs wouldn't respond, as if they were made of cotton."

When Viktor Briukhanov arrived at the plant, having been told only that there had been an explosion, and saw that the top of Unit 4 was missing, he knew immediately that his career was over, and that he would almost certainly end up in jail. "This is my prison," he thought. In the Soviet system, the director was inevitably held criminally liable for any disaster. Meanwhile, in the control room, shocked operators tried to make sense of confusing readings. They seemed to indicate that there was no coolant going to the reactor, while at the same time the emergency control rods had only partially engaged. Diatlov knew that this spelled disaster, irrespective of what had already gone wrong; the fuel rods were likely still fissioning away, and with no cooling the reactor would surely melt down (or at least, it would have done if it had not already exploded). After some futile efforts to get the control rods moving, Diatlov went to check out what was happening in the turbine hall, which someone had told him was on fire. The scene that greeted him reminded him of Dante's *Inferno*: "Streams of hot water were bursting in every direction from the damaged pipes and falling on the electrical equipment. There was steam everywhere. And the crackling sounds of short circuits in the electrical system resounded as sharply as gunshots." The ceiling had been staved in, and beyond the twisted and hanging girders, Diatlov could see the stars.

He helped engineers who were desperately trying to prevent a machine oil cache from catching fire, and encountered unfortunate workers who had suffered terrible steam burns from the first explosion. One man had been killed immediately, while another would soon die of his injuries. Observing the ruined reactor building, Diatlov and Trehub swapped appalled responses. "This is Hiroshima!" moaned Trehub. "I've never dreamed anything like this even in a nightmare," responded Diatlov. And yet they had not grasped the full extent of the disaster – like engineers and operators at Three Mile Island and many other disaster-struck reactors, they long resisted acknowledging that even their worst-case scenarios fell short of the mark. One man who might have informed them was engineer Aleksandr Yuvchenko, who had gone to Unit 4 to look at the reactor itself, only to find a mass of ruins and a vision of strange and terrible beauty: "I could see a huge beam of light flooding up from the reactor. It was like a laser light, caused by the ionisation of the air [by the intense radiation]. It was light-bluish, and very beautiful." Yuvchenko knew better than to linger for a few moments, but even that was enough to imperil his life, for he received a massive dose of radiation and was fortunate to survive, after months of treatment.

In the control room, there was awareness that radiation levels were rising, but in a defect that would characterize the entire response, the dosimeters available had such restricted ranges that the true exposure levels were off the scale. The dosimeters available in the control room measured radiation using fractions of roentgen units per second (a roentgen is a unit of ionizing electromagnetic radiation, such as gamma and X-rays, named for the discoverer of the latter). Note that the roentgen is a measure of exposure, not of absorption (for which rads or grays are used) or biological impact (rem or sievert), and so is not the most useful or informative unit, but the figures recorded at Chernobyl can still provide information about relative levels. The available dosimeters measured microroentgens per second, but only went up to 1,000 (equivalent to 3.6 roentgens/hour). In parts of the control room the readings were off the scale. Diatlov and his colleagues assumed that the exposure rate was about 5 roentgens/hour. The maximum exposure permitted to operators in an emergency was 25 roentgens, so Diatlov chose to believe that they were alright for the moment. What he had not reckoned with was the situation outside the control room. Anyone who had ventured out for any length of time was starting to display symptoms of acute radiation exposure: darkened skin (known as a "nuclear tan") and headaches.

Danger was everywhere; when one of the turbine engineers tried to search the engine room for a missing man, he had to climb through a cascade of water pouring out of a broken pipe. The water was radioactive, and the man would die from his exposure. Diatlov, who had accompanied him, was exposed to an estimated 325 roentgens of radiation, normally a fatal level, and suddenly began to feel the effects: nausea and extreme exhaustion. He was sent to talk to Briukhanov but was able to offer no explanation beyond a readout showing that the control rods had got stuck, and was then bundled into an ambulance.

The men Diatlov left in the Unit 4 control room still believed there was a reactor to save and were focused on trying to keep pumping emergency coolant into it. But by this time Briukhanov should have known better, because he had already dispatched his deputy chief engineer, Anatolii Sitnikov, on what would turn out to be a suicidal fact-finding mission. Sitnikov had climbed up on to the roof of the shattered reactor hall and looked down into the reactor, thus exposing himself to a lethal dose of radiation. "I think it's been destroyed," was his gloomy diagnosis, "It's blazing away." Yet even Sitnikov refused to engage with this reality, joining in with efforts to get more water into the reactor.

The dreadful reality proved similarly difficult to acknowledge for those encountering blocks of graphite scattered around the plant. Desperately they sought for an explanation other than the obvious; no one wanted to believe that the graphite had come from inside an active reactor. Engineer

Arkadii Uskov described the typical reaction, "What we saw was so terrifying that we were afraid to say it aloud." His diary entry for that day recorded, "The mind refuses to believe that the worst that could happen has happened." Sergei Parashin, a nuclear engineer who was the head of the Communist Party committee at the plant, later described the collective failure to acknowledge reality: "The stress was too great, and our belief that the reactor could not explode was also too great. Mass blindness. Many see what has happened but do not believe it."

This blindness would corrupt the entire official response to the disaster. Nobody wanted to admit the truth, not least because no one wanted to take responsibility – for the accident, for the response, or even for being the bearer of bad tidings to higher ups. This was the corrupting effect of an authoritarian regime, which punished honesty and encouraged buck-passing. It infected almost every aspect of what followed. To give just one example, when Briukhanov's team prepared a briefing for the regional Party committee, they listed the radiation levels at the plant as 1,000 microroentgens per second. Briukhanov knew that the levels were probably much higher, but since this was the maximum reading on the dosimeters, he was content to list this as the figure.

The true number was hair-raising. Serafim Vorobev, head of the Chernobyl civil defence department, had the only dosimeter with a range that went up to 200 roentgens. In the underground emergency bunker where the plant management were now gathering, Vorobev measured an exposure rate of 0.03 roentgens/hour, 600 times more than normal. Outside the bunker the level was five times higher, and near Unit 4 his dosimeter went off the scale, indicating a rate higher than 200 roentgens/hour. In fact, radiation was streaming out from the exposed reactor at a rate of 30,000 roentgens/hour, and any of the chunks of fuel and graphite scattered about was giving off 5,000 per hour. This was the radiation field that had bathed the unfortunate firefighters up on the rooftops, battling to put out the flames; six of them would not survive the doses of up to 2,000 rads (20 grays) – roughly equivalent to 2,000 rem – they were later estimated to have absorbed.

At the hospital, the relatives of the firefighters desperately fought for access to their loved ones, and were dismayed to find them swollen, puffed up and weak. The firemen told their families to get out of town. Pravyk told his parents to get his wife, wrap their newborn baby in blankets and flee to his parents-in-law in central Ukraine, far away. According to one story, his eyes had been burned from brown to blue by the intensity of radiation to which he had been exposed. The firemen would eventually be airlifted to Moscow, where six of them would die, along with 22 others who had suffered massive acute exposure on the first day of the disaster. They included two security guards standing sentinel near Unit 4, who stayed at their posts and received fatal doses as a reward.

Inside the command bunker, however, no one wanted to listen to Vorobev's bad news about dose rates. Parashin later said, "he was frightening to look at . . . I didn't believe him." Vorobev wanted Briukhanov to sound the sirens and warn the civilian population that there was a radiation danger, but the plant director would not listen, and told him to restrict his information to the civil defence HQ only. Like many of the plant's top brass, Briukhanov was in a state of despondency and near torpor, like a sleepwalker, stunned and depressed by the scale of the disaster and the certain knowledge that they would get the blame.

Gorbachev was informed about the accident just after 05:00, almost four hours after the explosion; the initial briefing, based on Briukhanov's misleading radiation report, was that the situation was not too serious. The Kremlin's response was to do what the Soviets always did: appoint a commission. Officials and nuclear experts from around the country were directed to go to Chernobyl. They began arriving in Kiev around lunchtime.

The most senior man on the spot initially was Anatolii Maiorets, the all-union minister of energy, who had flown in from Moscow, initially believing, like Gorbachev, that the incident would soon be cleared up. Even once he had been briefed that matters were much worse than he had been told, he was disinclined to listen. When a Ukrainian official, General Hennadii Berdov, told him that 1,100 buses had been mobilized to be ready to evacuate Prypiat,

Maiorets dressed him down. "What's all this talk of evacuation?" he barked, "Are you trying to start a panic?"

There was a general resistance to moves that would openly acknowledge the severity of the situation, and no one wanted to take responsibility for making things look bad, even if that meant denying reality. "I am opposed to evacuation," Maiorets insisted, "The danger is clearly exaggerated." But he was about to get a rude awakening. Two nuclear experts from Moscow, Boris Prushinsky and Konstantin Polushkin, had arrived in Prypiat that afternoon, and requisitioned a helicopter so that they could get a proper look at the accident site. It was all too obvious that the reactor had exploded and the situation was catastrophic. "The upper biological shield of the reactor was now a bright cherry-red from the extreme heat and was lying at an angle over the reactor vault," Prushinsky reported. "It was safe to say that the reactor was utterly finished." Maiorets asked what they ought to do. "God knows," Prushinsky told him. "Right now I can't say. There is graphite burning in the reactor. That has to be extinguished before anything else. But how and with what? We've got to think."

Another nuclear expert who had flown in, Valery Legasov, of the Kurchatov Institute (the leading nuclear research institution in the USSR), was to assume a leading role in the response. He was similarly shocked by the scale of the disaster, relating how, on arrival at the plant, he had been appalled to see "a crimson glow that expanded to fill half the sky. We could see a white pillar several hundred metres high consisting of burning products constantly flying from the reactor."

By around 20:00, after spending much of the evening arguing, the commission was coalescing around a plan to use helicopters to dump material on to the burning reactor to extinguish the blaze, but resistance to evacuation remained strong. Just after 21:00, however, came three more explosions. The nuclear experts were fearful that fuel in the stricken reactor might go supercritical, causing such an explosion, but it could have been a steam explosion caused by water falling on to the red-hot graphite. Whatever the cause, with the wind picking up, it was increasingly apparent that Prypiat was in escalating danger. After further debate and attempts to palm the problem off to superiors, the decision was finally taken in the early hours of 27 April. Prypiat would be evacuated. But what about Kiev, just 130 kilometres (80 miles) away? An order of this magnitude would mean admitting to the world that the Soviet nuclear industry was responsible for a calamity, news that the authorities were desperate to suppress. Radioactive gas and particles, however, were not at the behest of the Party, and already the telltale evidence of the disaster was drifting across borders and raining down all across northern Europe.

Probably the first people outside Chernobyl itself to know about the disaster were the US military, for it later emerged that an American spy satellite was passing over Ukraine at the time of the accident, and that the US military might have known of the incident before the Soviet government. They did not reveal the intelligence, however, in order to conceal their surveillance capabilities. Outside of the intelligence community, the Finns were probably the first to detect signs of the disaster. After instituting their own nuclear power programme, the Finns had installed a network of radiation detectors around their country to keep an eye out for leaks and contamination. Early on the morning of 26 April, a border patrol agent on the south-east border with the USSR, about 1,200 kilometres (745 miles) away from Chernobyl, noted that a radiation detector was indicating an alarm. The information was passed up the chain to the premier, but, wary of his nation's delicate relationship with the Soviets, he did not announce the information to the world.

It was left to the Swedes to break the news. On the morning of 27 April, about 30 hours after the explosion and a thousand miles away from Chernobyl, a radiation monitoring specialist noticed something strange at the Forsmark Nuclear Power Station, two hours from Stockholm. Cliff Robinson found that he kept setting off a radiation alarm. Initially dismissing it as a malfunction, his suspicions were aroused later that day when he noticed a line of workers with the same problem. Robinson ran tests on the bottom of a shoe and found that it was contaminated with nuclides not found at their plant – fission products that initially made him wonder if there had been a nuclear explosion. Cross-referencing with radiation monitors at other nuclear plants, he quickly realized that the contamination was blowing in from the south-east; presumably from the Soviet

Union. It was becoming apparent that a plume of radioactive contamination had spread north-west across Scandinavia, a plume that included radioactive xenon and krypton, as well as a stew of noxious fission products. In Finland, for instance, levels of gamma radiation were being measured at six times the norm.

Despite repeated requests by Sweden and the IAEA in Vienna, the Soviets denied any knowledge, but the contamination was clear. It would eventually lead to widespread precautionary measures, with prohibitions on picking mushrooms and berries, and entire herds of reindeer being culled and buried, although Scandinavian authorities later recognized that this may have been unnecessary. In the UK, some upland sheep pasture areas were so badly affected that lamb from the area was banned until 2012. By Tuesday, 29 April, the Chernobyl disaster was on the front pages of newspapers around the world, but not in the Soviet Union. The main Ukrainian newspaper relegated a report about an "incident" to the bottom corner of page three.

By this time, events in the Ukraine had moved on. With the decision made to evacuate Prypiat, things moved fast. At 13:00 on the 27th, the city's radio station transmitted instructions about the evacuation:

> Attention! Attention! In connection with the accident at the Chernobyl atomic power station, unfavourable radiation conditions are developing in the city of Prypiat. In order to ensure complete safety for

> residents, children first and foremost, it has become necessary to carry out a temporary evacuation of the city's residents . . .

Residents were given just two hours to prepare, before most of them were loaded into a fleet of 1,125 buses. They were falsely told that they needed to bring only enough for three days. Prypiat resident, journalist Liubov Kovalevskaia, recalled,

> Everyone was dressed as if to go camping, people were joking, and everything was rather quiet all around. There was a policeman beside every bus, checking residents according to a list, helping people bring in their belongings, and probably thinking of his family, whom he had not even managed to see in the course of those twenty-four hours.

Earlier in the day, an amateur film crew had been shooting footage of people enjoying the weekend break, and of preparations for the May Day fair. Now they captured on film the evacuation, along with a chilling token of the otherwise invisible menace driving them from their homes: white streaks on the film, caused by radiation from particles settling on the city. People were not permitted to bring their pets, and later a special squad would be organized to shoot the dogs left roaming the streets of the abandoned city. In all, 44,460 people were evacuated. Nearly 5,000 plant workers remained behind to operate the plant and oversee the shutdown of the other reactors, as well as elderly people who refused to leave.

Just 12 kilometres (7 miles) away from the plant, however, the town of Chernobyl was not evacuated for another week, when the authorities declared a 30 kilometre (18 mile) exclusion zone around the plant. Controversy raged among officials over what to do about Kiev, where parades and outdoor events were planned to mark the 1 May holiday. Even when the wind changed to blow radionuclides towards Kiev, the authorities refused to cancel the marches, let alone evacuate. An invitation went out inviting schoolchildren to attend parades; the children of the political elite had already been flown out. By 14 May, some 116,000 people who had been living within the zone had been evacuated, and later relocated. But the exclusion zone failed to take proper account of the actual pattern of fallout, with the result that heavily contaminated villages in Belarus were not evacuated for months after the accident. Indeed, evacuations continued sporadically for some time, and in the years following the accident, another 220,000 people were resettled into less contaminated areas, while the initial exclusion zone, which had spanned an area of 2,800 square kilometres (1081 square miles), was extended to cover 4,300 square kilometres (1,660 square miles).

Once the solution of helicopters dumping extinguishing materials had been adopted, it was pursued with a will. On the morning of 27 April, with no one yet requisitioned to do the job, a general and two powerful officials found themselves dragooned into filling sandbags. All the while the deputy head of the Moscow government and newly arrived head of the emergency management committee, Boris Shcherbina, berated them, screaming at them that they "were very good at blowing up reactors but useless when it came to filling sandbags".

The first attempts to drop the sandbags were desperate and dangerous. A helicopter would hover over the burning reactor and a crew member would toss sandbags from the door, attempting to aim them through the gap between the skewed upper biological shield and the shattered reactor sides. Not only was this hopelessly ineffective, it exposed the helicopter crews to dangerous levels of radiation. Soon the pilots and other crew refined the operation, establishing a lookout point to direct operations, and rigging up a hook mechanism so that they could release large loads from safer altitudes. Over the next 10 days, 1,800 helicopter flights dumped about 5,000 tonnes of material onto the reactor, including 1,800 tonnes of sand and clay, 40 tonnes of boron compounds, 2,400 tonnes of lead and 600 tonnes of dolomite. The sand and clay was supposed to smother the fire, block radiation and limit the release of particulates; the neutron-absorbing boron compounds were to limit the chances of a critical excursion and shut down any fission; dolomite was added to act as a heat sink, and to release carbon dioxide to help choke the fire; and the lead was there to block radiation.

Unfortunately, this heroic effort not only proved largely ineffective – most of the material missed its target and instead piled up on the floor to the side of the reactor pit – but was probably counter-productive. It is now believed that it formed an insulating layer, trapping heat in the melted remains of the reactor, so that while it quieted at first, it was still bubbling away, ready to flare up again a week later and trigger a further massive release of radionuclides.

For the nuclear experts on site, Legasov prominent among them, the real concern was the "China Syndrome" (see page 240). They feared that whatever fuel was left in the reactor would melt its way through the lower structures and plunge into the reserve coolant, aka "bubbler", pools, which had foolishly been designed into the building around the base of the reactor. These contained up to 20,000 tonnes of water, now badly contaminated because of the initial and woefully misguided attempts to pump up to 300 tonnes of water per hour into the reactor, when it was still believed that it was intact. Contact between melted core and water would trigger another steam explosion, potentially worse than the initial ones. Beyond this there was the horrible scenario that the core might melt right down to the water table, and from there spread out to poison the entire ecosystem of the region.

The first priority was to drain the bubbler pools by opening the valves, but the passages to the valves had been flooded by the initial deluge, and the water was highly radioactive.

Wading and swimming through it to open the valves was probably a suicide mission. With special financial and material incentives offered by the authorities, three engineers volunteered for the task, although at least one later disputed this, saying, "I was ordered to go there, so I went". On 2 May, rigged up in makeshift diving-cum-radiation suits, the three brave men ventured into the deadly basement and opened the valves of the bubbler pools, letting out the water into adjacent chambers so that it could be pumped clear. Legasov remembered the conflicting emotions of one of the divers when he was presented with his cash reward:

> . . . he found it inconvenient to refuse the money, but at the same time the monetary form of the award itself gave him little satisfaction because in fact, and particularly at that time, people were struggling to deal with the accident, to bend all their efforts and do all they could, not thinking of any incentives, whether material or moral.

The divers were not expected to survive, and indeed it is often reported that they all died shortly afterwards, but in fact all three of them survived.

Although the risk of another steam explosion could now be reduced, there still remained the threat of a China Syndrome. In the base of the ruined reactor, a bizarre, almost alchemical stew of reactor core elements had combined into lava-like fuel-containing material (LFCM), later called "corium" (because it came from the core), a profoundly radioactive lava of molten fuel, concrete, graphite, metal and other elements. At the very moment that the divers were negotiating the flooded corridors in the basement, not far above them the corium was starting to burn through the reactor floor. Over the next few days nearly 100 tonnes of corium would ooze down pipes and into the basement, eventually solidifying into a range of strange forms, from stringy stalactites to a grotesque appendage later dubbed the Elephant's Foot.

The possibility of such a meltdown prompted the concoction of a bold plan to try to cool the base of the blazing reactor, allowing time to install some sort of containment shield or platform beneath it. The freezing was to be achieved by injecting into the ground beneath the reactor colossal quantities of liquid nitrogen – up to 25 tonnes per day of the fluid at a temperature as low as –195°C (395°F). The process began on 4 May but soon proved impractical and was scrapped. Instead work began on a cooling and containment platform: starting from under Unit 3, miners and metro workers tunnelled beneath the reactor building, so that a massive reinforced concrete slab with a built-in cooling system could be installed. It took 400 workers 15 days to complete the tunnels.

From 4 May onwards, with these interventions in hand, it seemed like the tide was turning. Over the next few days the radioactive output of the stricken and burning reactor fell; on 6 May it was estimated at 150,000 curies, 100 times lower than the day before. By 8 May, 20,000 tonnes of water had been pumped out from beneath the reactor. Legasov was confident that the worst was over. On 9 May, he and others prepared to celebrate Victory in Europe Day, but at the reactor there was a resurgence of the fire. Flames leapt up into the air and a pink glow appeared. Legasov recalled in his memoir: "We were embittered, of course. The May 9 holiday was ruined."

The cause of the flare-up was not clear. The general opinion is that the graphite fire was already substantially extinguished by this point. Possibly this new blast of emissions was due to corium flows reaching some residual water, causing a steam explosion, or a collapse of part of the hardened crust of dumped material and melted reactor parts, which had formed over the still baking core remnants. This same day, 9 May, the firefighter Lieutenant Pravyk died from his ARS.

According to the IAEA, the fire was definitely out by 10 May. Operations now moved into the containment and clean-up phase. Hundreds of thousands of people were mobilized from all across the Soviet Union to "eliminate the consequences of the Chernobyl nuclear plant accident". An estimated 500,000 liquidators (*likvidatory* – see page 71) would eventually be called up, mainly military reservists and conscripts but also prisoners and others. Most of the conscripts were between 18 and 20 years of age. The regime showed a chilling disregard for their safety and health; they were largely considered to be expendable. This applied above all to the "biorobots": the personnel detailed to clear up radioactive debris, often by hand with only the most perfunctory and makeshift protective equipment. Draftees from the Central Asian Republics, such as Turkmenistan and Uzbekistan, whose Russian was shaky and who knew almost nothing about the risks and realities of radiation, were the favoured biorobots.

Over the following weeks, liquidators would clear reactor debris from the site and seek to reduce contamination in the surrounding region, spraying sticky substances known variously as "water-soup" and *bourda* ("molasses"), with hoses and from helicopters, on to roads and buildings, soil and vegetation. These viscous substances were supposed to trap radioactive dust, so that it could be marshalled and eventually buried. Entire villages were bulldozed and buried. One of the most famous of the liquidators' tasks was the felling and burying of the Red Forest, a 10 square-kilometre (4 square-mile) stand of pine trees which had absorbed so much radiation that they had turned red. Another notorious episode was the clearance of the roof of Unit 3. The three surviving reactors at the site had been shut down within 24 hours of the accident. Officials were desperate to get them back online, but Unit 3 was heavily contaminated with chunks of reactor debris on the roof. A military contingent of biorobots, 3,000 strong, cleared the debris by taking turns to scoop up a piece of debris with a shovel, run to the edge of the roof and drop the debris over the side and then sprint back inside the building.

On 20 May, design work began on a colossal containment structure, officially designated the *Obyekt Ukrytiye*, or "Object Shelter", but generally known as the sarcophagus. The idea was, following on from the completion of the cooling slab, to enclose the still highly radioactive ruins of the reactor inside a giant concrete coffin. Construction would begin in June and continue for 206 days, finishing in late November and involving 200,000 workers. Vast quantities of material were put in place in punishingly dangerous conditions: over 400,000 tonnes of concrete, enough to fill more than a third of the Empire State building; and 7,300 tonnes of metal framework. The immense construction would eventually enclose 750,000 cubic metres (26,500,000 cubic feet) of heavily contaminated debris and soil.

The sarcophagus was not intended to be a permanent solution, and over the decades it suffered serious deterioration, with extensive water leakage. Intended to stabilize the sarcophagus via a series of cantilevers, a massive, 63 metre/207 foot-tall cat's cradle of yellow metal work, the Designed Stabilization Steel Structure, was added to the outside. In 1996, radiation levels inside the sarcophagus were still at extreme levels: up to 10,000 roentgens per hour (compared with a typical background level of 20–50 microroentgens/hour). Given that a lethal dose is estimated at 500 roentgens over 5 hours, it was clearly going to be impossible to repair the sarcophagus properly. A competition had been launched in 1992 to design a new structure, and the winning submission was for a sliding arch, which could be constructed away from the high radiation zone and then slid into place. Known as the New Safe Containment (NSC), this giant arch was made from a steel skeleton around two immense concrete beams, supporting a skin of cladding panels made from a sandwich of materials. Its dimensions are enormous: externally it is 108 metres (354 feet) high and 162 metres (531 feet) long, with a span of 257 metres (843 feet), and internally it has a clearance of 92.5 metres (303 feet) and a span of 245 metres (804 feet). This makes it big enough to house the Statue of Liberty, St Paul's Cathedral or Notre Dame, depending on whether you like your comparisons with New York, London or Paris. At 36,000 tonnes, the NSC weighs more than the Eiffel Tower. It is fire- and tornado-proof and designed to last for at least 100 years. Although it was originally scheduled to be complete by 2005, it was not until 2017 that the NSC was actually manoeuvred into position.

With the sarcophagus in place, the immediate clean-up operation complete, and over 300,000 people relocated, attention turned to the long-term impact of the disaster. With a rating of 7 on the INES scale, Chernobyl remains the worst nuclear disaster in history. Roughly a third of the contents of the reactor were released into the environment, although estimates of exactly how many curies vary, as do the comparisons and equivalents that stem from this. The extent of the release is estimated at anywhere between 50 million and 200 million curies. The typical comparison is with Hiroshima, but such comparisons are problematic because the explosion at Hiroshima was an airburst, which did not generate that much fallout. Chernobyl is said to have equalled anything from 10 to 500 Hiroshimas, as well as 75 Kyshtym disasters and 3,000 Windscales.

The most worrying radionuclides released were the ones with intermediate half-lives – long enough to hang around in the environment for decades, and short enough to cause dangerous levels of radioactivity. Particularly concerning were strontium-90 and caesium-137, with half lives of 28.8 years and 30.07 years respectively. The former, being chemically similar to calcium, is readily taken up by the human body and incorporated into bones, while the latter is similar to potassium and thus also taken up by the body. Perhaps the greatest long-term health effects were caused by the 45 million curies of radioactive iodine emitted in the accident: at least a thousand times more than were released at Windscale.

Although the sheer scale of the release was calamitous, comparison with the contamination produced by nuclear testing and early plutonium production offers some perspective. Early Soviet and American plutonium plants are estimated to have generated 200 million curies of contamination, while the 520 weapons detonated by various nations in atmospheric testing, created billions of curies of radionuclides. American and Soviet bomb tests are estimated, for instance, to have generated at least 20 billion curies of radioactive iodine between 1945 and 1962.

The long- and short-term health consequences of Chernobyl are heavily contested. It is not even clear how many people died of ARS. Originally 237 people were diagnosed with ARS, as a result of exposure during the accident, firefighting and clean-up, but this was later revised down to 134 cases. A figure of 50 deaths from ARS is often given, but this is disputed. Within a few weeks of the accident, 29 people died as a result of ARS, although the figure is sometimes given as 28. Between 1987 and 2004, 19 more workers died, but attributing their deaths to radiation exposure is not straightforward.

Still greater uncertainty surrounds the fate of the liquidators. The average dose to which liquidators were exposed was 120 millisieverts (mSv), although about 20,000 received doses in excess of 250 mSv, according to a 2008 report by the UN Scientific Committee on the Effects of Atomic Radiation (UNSCEAR). The same report estimated that 150,000 civilians received doses over 50 mSv, of whom around 6,000 received over 100 mSv, the threshold for elevated health risks. What impact did this have? A Kiev museum exhibit about the disaster claims that by the 2010s, 45 per cent of liquidators had died and 50 per cent developed disabilities. But the truth is that there was very little systematic monitoring or follow-up. Leonid Ilyin, a former member of the International Commission on Radiological Protection, told science journalist Fred Pearce: "None of these men were registered by name. None was checked on a regular basis. They all went back to their homes." A WHO estimate was that 2,000 of the liquidators would eventually die of radiation-linked illness.

Contamination from Chernobyl spread all across northern Europe. The worst hit countries were Ukraine, where 5 per cent of its territory (38,000 square kilometres/14,672 square miles), inhabited by 5 per cent of its population, was contaminated; Belarus, where the corresponding figures were 23 per cent (44,000 square kilometres/16,990 square miles) and 19 per cent; and Russia, with 1.5 per cent (60,000 square kilometres/23,166 square miles) and 1 per cent. At the time there were alarmist predictions about the potential impact on long-term health, with suggestions that up to a million lives would be shortened by increased cancer risk and other effects. Even more sober assessments were scary: in 2005 UN agencies put the eventual toll at 4,000, while Greenpeace International put it at 90,000.

Since then, however, several studies have suggested that the long-term health impact has been minimal. In 2018, UNSCEAR concluded that, "There is no scientific evidence of a major public health impact attributable to radiation exposure . . . apart from a high level of thyroid cancer in children." They estimated that Chernobyl caused 5,000 extra thyroid cancers, of which, at most, around 1 per cent – 50 people – would die prematurely as a result.

Some experts have tried to put the excess mortality burden of Chernobyl in context. Quoted in Richard Rhodes'

1993 book, *Nuclear Renewal*, Bernard Cohen, a professor of physics and radiation health at the University of Pittsburgh, estimated that, "The sum of [Chernobyl-related] exposures to people all over the world will eventually, after about fifty years . . . cause about 16,000 deaths." At the time this was the same as the annual number of deaths caused by air pollution from coal-burning power plants in the United States alone, while air pollution in general killed 100,000 Americans a year. "Everything we do involves risk," reflected Cohen, "Risk is an unavoidable part of our everyday lives."

The psychological impact of the disaster was, however, indisputable. The IAEA's International Chernobyl Project reported that among local populations there were high levels of "anxiety, depression and various psychosomatic disorders attributable to mental distress", and spoke of a "paralysing fatalism". Millions of people in Ukraine and Belarus believed that they had been affected, with anxiety "out of all proportion" to actual risks, which was "extremely harmful to people". Large numbers of women were incorrectly advised by their doctors that their risk of birth defects was so great that they should consider abortion, and one physician who helped to treat victims of the disaster, Professor Robert Gale, writing in the journal *The Cancer Letter*, said, "We estimate incorrect advice from physicians regarding the relationship between maternal radiation exposure from Chernobyl and birth defects resulted in more than 1 million unnecessary abortions in the Soviet Union and Europe. Ignorance is dangerous." Another sad victim of Chernobyl was Legasov, the nuclear expert from the Kurchatov Institute, who was heavily criticized, both for his role in the original design of the RBMK reactors and for speaking out too frankly about the causes of the accident. Two years to the day after he arrived in Chernobyl, he hanged himself in the stairwell of his apartment block.

As Briukhanov had predicted, he and other senior managers at Chernobyl had to carry the can for the disaster. On 3 July 1986, he was sentenced to 10 years in prison for "serious errors and shortcomings in the work that lead to the accident with severe consequences". His health crippled by radiation exposure, he was released in 1991 after serving half his sentence. But the typical obfuscations of the Communist regime could not protect them from the dire political fallout of a disaster that was made in the USSR, a truth that was all too apparent. "As members of a select scientific panel convened immediately after the . . . accident," wrote American physicist and Nobel laureate Hans Bethe, "my colleagues and I established that the Chernobyl disaster tells us about the deficiencies of the Soviet political and administrative system rather than about problems with nuclear power."

In Svetlana Alexievich's 1997 landmark *Voices from Chernobyl*, historian Aleksandr Revalskiy observed, "Chernobyl is the catastrophe of the Russian mind-set. It wasn't just a reactor that exploded, but an entire system of values." Anger about the disaster and how it was handled led to the rapid growth of a political movement in the Ukraine that mixed green concerns with nationalist ones and helped boost the cause of Ukrainian independence. This in turn helped to bring about the break-up of the USSR.

As for the power plant itself, it was back in business remarkably quickly after the accident. Unit 1 was restarted on 29 September 1986 and connected to the grid two days later, and on 10 October construction work on Units 5 and 6 resumed. The Chernobyl power plant would continue to operate until 2000, when it was finally shut down, in the face of protests by the plant workers. In 2017 a solar power farm opened next to the site of the plant.

Today the exclusion zone is a popular tourist destination, and also seems to have become a haven for wildlife. According to one study, there has been a seven-fold increase in the population of large mammals, although other studies have found that small mammals and birds do demonstrate a health burden from higher radiation exposure. Over half of the exclusion zone is forest. Although Soviet decommissioning efforts in some places involved bulldozing areas, stripping away all the forest and topsoil, and burying and/or concreting over everything, it was cheaper on the whole to leave the forest intact, and use it to lock away many of the most dangerous radionuclides in the wood, leaf litter and soil. But the risk of this approach is that forest fires, which are all too common, will release large quantities of dangerous isotopes. For instance, in April 2020, around 100 hectares (247 acres) of forest burned. Radiation levels were said to have spiked at 16 times above normal, according to Yegor Firsov, head of Ukraine's state ecological inspection service, although he later reassured worried residents of Kiev that they "didn't have to be afraid" of opening their windows.

Some 7,000 workers are employed decommissioning the plant, and although most commute from outside the exclusion zone, many spend periods of up to a fortnight there. Even Prypiat itself is not entirely abandoned. By the 1990s, around 1,800 people, known as self-settlers, had illegally moved back into the exclusion zone, to resume their lives there; there was even said to be a community of self-settler monks. This number has probably dwindled now, but in 2013 at least 200 self-settlers remained. In 2010, the Belarussian Government announced that it would seek to resettle much of the evacuated area, with tens of thousands of evacuees or their descendants. How safe is it to live there? Apart from in the immediate vicinity of the reactor (within a half-kilometre radius), radiation levels in the exclusion zone fell rapidly, so that average doses after a few years were less than 50 per cent above a normal background of 2.5 mSv/year.

This debate is emblematic of a wider one within the nuclear science and radiobiology community: what is a safe dose of radiation? Some argue that there is no such thing, while others seek to put dose rates in the context of natural background levels, likely voluntary exposures through phenomena such as X-rays and air travel (see Appendix), and other sources of risk. Similarly, the biological and ecological legacy of Chernobyl is heavily contested. Non-nuclear power plant accidents have, in the past, caused far greater loss of life. The collapse of the Banqiao Dam in China, in 1975, may have killed over 100,000 people. In August 2009, 75 people were killed when a turbine failed catastrophically at the Sayano-Shushenskaya hydroelectric power station in Russia. Whether Chernobyl should be assessed differently to such disasters, depends to a large extent on one's attitude to nuclear power in general. What makes nuclear disasters different is that their consequences may be uniquely long-lasting. Plutonium-239 does not exist in nature, but it was produced in Unit 4, and subsequently blasted across the nearby countryside. It has a half-life of 24,000 years, and since radioactive contaminants are dangerous for 10 to 20 times the length of their half-lives, the plutonium released at Chernobyl will continue to pose a threat for the next half a million years. If you are unlucky enough to ingest a single particle of plutonium dust, it could be fatal. The alpha radiation emitted by plutonium inside the body is between 10 and 1,000 times more damaging to chromosomes than the beta or gamma radiation emitted by radionuclides such as iodine, strontium or caesium. But the heavy plutonium particles spread around Chernobyl to a radius of only around 4 kilometres (2½ miles). Epochal ecological calamity, or localized inconvenience? Chernobyl's plutonium footprint seems to offer a prism through which to view its wider legacy.

Chapter 10

CONCENTRATION CRITICAL: TOKAIMURA, 1999

THE ACCIDENT at the Tokaimura nuclear fuel processing plant on 30 September 1999, resulted in the deaths of two workers. The extraordinarily awful and lingering death of one of them, Hisashi Ouchi, is one of the darkest episodes in the history of nuclear science, with accusations that physicians deliberately drew out his inevitable end as part of a ghoulish investigation into the effects of radiation. The incident also threatened to disperse radioactive material into the environment, spreading fear and confusion in the local community, and was widely perceived as a wake-up call for Japan's nuclear industry.

The Tokaimura incident was the result of poorly trained and unsupervised workers mixing the wrong quantities of uranium in the wrong way in the wrong container. The consequences might initially have appeared low-key: a wispish blue glow over the mixture was the only immediately apparent sign that something had gone horribly wrong – but by the time Ouchi saw it, his fate was already sealed.

This fatal mix-up occurred at a small nuclear fuel preparation plant operated by Japan Nuclear Fuel Conversion Co. (JCO), a subsidiary of Sumitomo Metal Mining Co. Unlike similar plants in Western countries or Russia, which would be expected to be placed as far away from urban centres as possible, this plant was in the middle of a city of hundreds of thousands of people: Tokaimura, in Ibaraki Prefecture. Since the plant was not actually running a reactor or handling waste, its processes were conducted partly beyond the supervisory scope of the normal atomic energy regulatory apparatus, and standards were not as stringent. Normally the plant processed low-enriched uranium fuel, but in a special division of the plant, the Fuel Conversion Test Building (FCTB), workers processed highly enriched fuel destined for Japan's experimental Jōyō fast breeder reactor. In contrast to the normal 5 per cent enriched fuel, the FCTB would process fuel with U-235 concentrations of up to 20 per cent, and it would have been expected that the plant's licensing conditions would be tightened accordingly. Yet Japanese regulators, perhaps through some combination of over-confidence in the discipline of Japanese workers and misplaced faith in the robustness of JCO's procedures, considered that the risk of a criticality incident did not meet the criteria for a "credible accident". That is to say, it was deemed vanishingly unlikely that such a concatenation of mistakes could occur as to cause such an incident. As a result, the FCTB plant was not required to install criticality alarms. Misplaced faith in "credible accident" criteria would come back to haunt the Japanese nuclear industry in the Fukushima disaster (see page 322).

The fuel process undertaken in the FCTB was surprisingly low-tech. First, a powdered uranium compound was mixed and dissolved in nitric acid, to give a solution of uranyl nitrate. This was then mixed further and transferred to a precipitation tank, where it reacted with ammonia to precipitate out uranium oxide, which could be collected in a subsequent stage. In the original licensed process, the potential for the mixture to go critical was limited by design elements such as the use of different, specifically shaped tanks for dissolution, mixing-and-storage and precipitation. The geometry of such tanks can help to prevent a critical mass of uranium collecting together (see Appendix). Specifically, the dissolution tank handled only small amounts of uranium at a time, and then transferred the solution, via a narrow pipe, to a long, thin holding vessel, from which it was then supposed to be drained into 4 litre (8½-pint) plastic bottles – too small to allow a critical mass to gather. The final stage was to transfer controlled amounts of the solution to a precipitation tank. Because it was never envisaged that this tank would hold significant quantities of uranium, there was no attempt to give it a criticality-proof geometry; quite the contrary, in order to save metal it was a circular tank, 45 centimetres (17¾ inches) across and 66 centimetres (26 inches) high. What is more, in order to remove heat generated by the chemical reactions of the material it was intended to hold, the tank had a water-cooling jacket around it. A vessel with a low surface area to volume ratio, in a jacket of neutron-reflecting water, which is filled with neutron-moderating aqueous and organic solutions, makes a pretty good design for a nuclear reactor, but no one at the JCO had put the pieces together to consider this possibility.

As long as the originally licensed procedures were meticulously observed, there would not have been a problem. In 1996, however, without permission from regulatory bodies, JCO dispensed with some of the tanks and began to let workers mix the powder and acid in steel buckets. To speed things up further, these buckets would then be tipped directly into the precipitation tank, which had impellers that could help mix the solution and accelerate dissolution. However, this meant bypassing the intermediate mixing-and-storage tank and made it much easier carelessly to circumvent controls over how much uranium was fed into the precipitation tank.

In September 1999, JCO won the contract to process 16.8 kilograms (37 pounds) of uranium for the Jōyō reactor, and on 29 September, Hisashi Ouchi, Masato Shinohara and Yutaka Yokokawa got started on the surprisingly unsophisticated manual process. They did not have adequate training, and evidently had limited understanding of the dangers of criticality, or of the potency of the fuel they were handling. The three colleagues were supposed to process the uranium in seven batches, each of 2.4 kilograms (5 pounds 4 ounces), to be dissolved into 5 litres (10½ pints) of nitric acid, and if they had restricted themselves to dealing with each batch separately, everything would have been fine. Unfortunately, all their previous experience was in processing 5 per cent enriched uranium. This time they would be handling 18.8 per cent enriched uranium.

Their immediate problem was that, on the favourable geometry holding vessel, the tap for draining the contents had too little clearance to fit the 4 litre (8½ pint) bottles easily underneath it into which the solution was supposed to be transferred. Accordingly, in line with the unsanctioned revised procedures, they decided to dispense with the holding vessel altogether. Instead, they started mixing up batches of uranium oxide powder and nitric acid in steel buckets, transferring the contents to large flasks and then emptying these through a glass funnel into the precipitation tank. Rather than processing the uranium in seven separate batches, as specified, they decided it would be much quicker to do it all at once; after all, the tank had a capacity of 100 litres (26 US gallons), and they would be making up only 45 litres (12 US gallons) of solution. With 5 per cent enriched uranium, this would not have been a problem. On the first day they dissolved four batches and emptied them into the precipitation tank. The next day they came back and got to work on the final three.

By 10:35 on Thursday, 30 September, they had successfully emptied in two more batches of solution, and Ouchi and Shinohara were tipping in the third and last. Ouchi was standing on a step next to the tank, holding the glass funnel, while Shinohara was at the top of a short ladder, holding the flask, which he was emptying into the funnel. Yokokawa was seated a few feet away, on the other side of a wall, doing paperwork at a desk. As Shinohara poured out the last few hundred millilitres, the mass of uranium in the tank increased to more than 16 kilograms (35 pounds), and the critical threshold was reached. The nuclear fission chain reaction within the tank became self-sustaining and a huge blast of neutrons and gamma rays radiated outwards.

The powerfully energetic radiation ripped apart air molecules immediately above the tank, creating a plasma of reactive ions that glowed with a ghostly blue light (see page xx). Yokokawa later reported that there had been a loud banging noise at the same instant; protected by distance and a wall, he received a radiation dose of 3,000–5,000 milliSieverts (mSv). Shinohara received a dose of up to 10,000 mSv (1,000 rem). Ouchi received a full body radiation dose that may have been as high as 20,000 mSv (2,000 rem). A dose of 5,000 mSv is usually fatal. For comparison, Cecil Kelley is estimated to have received a dose of about 120,000 mSv, and Louis Slotin about 21,000 mSv.

Ouchi and Shinohara staggered down from the steps and immediately started to feel the symptoms of massive acute radiation exposure, with disorientation, abdominal pain, nausea and breathing difficulty. Ouchi in particular would go on to display a similar progress of symptoms to Cecil Kelley (see page 134), losing control of his limbs and becoming unable to speak. Yokokawa had no clue what had just happened, but he heard alarms begin to sound;

they had been set off by the intense gamma radiation. He and Shinohara helped Ouchi stagger out of the FCTB, joining the workers piling out of the other facilities, one of whom noticed that the three of them looked in bad shape and called an ambulance to take them to the hospital.

Meanwhile, inside the precipitation tank, the uranium mixture was boiling violently; as vapour bubbles formed within it, so the concentration would dip below criticality and the nuclear reaction would abate, only to restart as the solution stopped boiling. Radioactive vapour was boiling off the top of the tank; might there be a risk of it leaking from the unsealed building?

As alarms continued to sound and gamma readings continued to show a potent source of radiation somewhere in the plant, supervisors traced the source of the readings to the FCTB. From the intensity of the radiation, it appeared that there was an unshielded nuclear reactor running somewhere within. Forty minutes after the accident, the company contacted Japan's Science and Technology Agency (STA), and by 11:30 the Ibaraki Prefecture (i.e. the local provincial government) had been told. An hour after this, the Prime Minister, Keizo Obuchi, learned of the accident. The residents of Tokaimura, some living just a few dozen metres from the accidental reactor, learned that something was afoot only when police began to cordon off roads within a radius of 200 metres (220 yards) of the plant at around 13:40. It was over an hour more until they were actually told anything, when the mayor issued an evacuation advisory to

161 people from 39 households living within a 350 metre (380 yard) radius of the JCO plant. Some hours later this was followed by a government order for those living within a 12 kilometre (7½ mile) radius, to stay indoors and keep their windows shut.

The fear among the authorities was that fission products boiling off the tank would escape the FCTB and settle on the surrounding community. In fact, relatively little radioactive material escaped from the plant; the FCTB was maintained at negative pressure (where the air pressure inside the building is lower than outside, so that air flows only into the building and not out of it) and air escaped only through stacks fitted with particulate filters. In the event, only noble gases and a tiny amount of radioactive iodine were released. Eventually ventilation from the building was switched off so that no gas would be able to leave. The greater risk to the houses immediately surrounding the plant came from the neutrons and gamma rays streaming from the impromptu reactor. The priority was to stop the reaction, but how? Although the power levels were gradually reducing, so that after 17 hours they had halved, there was still no way anyone could enter the processing room. Instead, the company hatched a plan to shut down the accidental reactor by draining water from the cooling jacket; without its reflecting jacket, the reactivity of the mixture in the tank would drop below the threshold for criticality.

Unfortunately, the apparently straightforward task of draining water from the jacket proved to be extremely technically challenging, involving workers approaching the outside of the FCTB for short periods at a time (to limit their radiation exposure) to dismantle pipes and open a valve. Eventually it was necessary to pump argon gas into the system to displace the last of the water, after which the jacket was filled with fluid containing boron, which soaks up neutrons. It would eventually take 20 hours from the start of the incident to put a stop to the critical reaction, and in the process 27 workers were exposed to radioactivity. This did not mean, however, that the plant was safe. Fission products inside the tank continued to pump out gamma radiation, and it was necessary to install shielding to protect people outside the building. The residents who had been evacuated from the immediate vicinity of the plant were only allowed to return home two days later, to find their houses shielded from gamma rays by sandbags.

It was later estimated that the accident had involved 2.5×10^{18} fission events, releasing 81 megajoules of energy (equivalent to burning 2.5 litres/5 pints of gasoline). Although 160 terabecquerels (TBq) of noble gases and 2 TBq of gaseous iodine were released, almost none escaped from the building. When the reaction was finally brought under control and shielding was put in place, radiation levels beyond the JCO site returned to normal. The International Atomic Energy Agency (IAEA) initially rated the incident as level 4 on the INES scale: "an event without significant off-site risk". This was despite the fact that members of the public were exposed to radiation; one was estimated to have received

24 mSv, four 10–15 mSv, and fifteen 5–10 mSv. In addition, 24 JCO workers received up to 48 mSv, and another 56 were exposed to doses of up to 23 mSv. Although none of these exposures exceeded the nuclear industry's maximum allowable annual dose of 50 mSv, 119 people were exposed to doses exceeding the 1 mSv level recognized by the International Commission on Radiological Protection as the maximum that members of the public can safely be exposed to in a year. Eventually the incident was rated as a level 5.

If the impact on the local area was, fortunately, negligible, for the three men at the centre of the storm it was life-threatening. After initially being helicoptered to the National Institute of Radiological Sciences (NIRS) in Chiba, just east of Tokyo, the men were taken a few days later to the University of Tokyo Hospital. As with other cases of acute – and eventually fatal – radiation sickness, the initial crisis was followed by a surreal few days in which the patients rallied. Coherent and relatively comfortable, they were able to laugh and joke, and even to talk about going home. But it was already apparent to most observers that Ouchi and Shinohara were dead men walking. For instance, a report in the *Irish Times* dated Saturday, 2 October (just two days after the incident), observed that "While they continue to receive treatment, there is really nothing that can be done to reverse the effects of a large radiation dose."

Yokokawa was the only one of the three to survive, leaving hospital on 20 December 1999. Hisashi Ouchi and

Masato Shinohara endured unspeakably grotesque, agonizing and drawn-out deaths that raised serious questions about the ethics of their treatment and the motivations of the physicians providing it. Ouchi in particular has gone down in history as the worst nuclear casualty the world has seen, and was the subject of a celebrated documentary and subsequent book, *A Slow Death: 83 Days of Radiation Sickness* (2008). Shinohara had absorbed a smaller dose, and thus survived longer, but as the less "superlative" case, his demise was not as well documented and does not seem to have been quite as gruesome. Ouchi, on the other hand, who was deemed to have absorbed as much radiation as someone who had been standing at ground zero at Hiroshima, and was fated to be Japan's first civil nuclear power victim, was the object of grim fascination from both the scientific/medical community, and the wider public.

At the NIRS, blood tests on Ouchi (35 years old) and Shinohara (40 years old), showed that their white blood cell counts had plunged to the point where they were effectively undetectable. Ouchi's bone marrow was liquefied, and chromosomes in his few remaining white blood cells appeared to have been completely shattered. Three days after the accident, doctors attempted the first of several novel and/or experimental procedures. First, the two men had blood transfusions, and later Ouchi was given stem cells from a sibling, while Shinohara received a transfusion from umbilical cord blood (also a source of stem cells). On 8 October, Dr Robert Gale of the University of California at

Los Angeles Medical Center, who had treated some of the victims of the Chernobyl disaster, arrived in Japan, at the request of Tokyo University Hospital. He described Ouchi's case as one of the worst he had ever seen, commenting, "it is going to require a lot of work and a lot of good luck to get him through all of the problems he is going to face in the next few weeks". The day after, David Kyd, a spokesman for the IAEA, based in Vienna, admitted that the chances of the two men surviving were slim. Both experts probably knew that even these qualified comments were optimistic, but physicians persisted in treating the two with increasingly desperate measures.

Meanwhile Ouchi's body began to fall apart. His skin sloughed off and his muscles liquefied. Blood leaked from his eyes like tears. After three weeks his intestines began to haemorrhage. Every organ in his body failed and he was kept alive by a battery of machines. He lost blood and other fluids at such a rate that doctors had to pump up to 10 blood transfusions into him every 12 hours. By day 63 of his ordeal he was receiving up to 10 litres (20 pints) of fluid daily, and eventually 20 litres (42 pints) a day had to be pumped into him because his bodily fluids were seeping out at such a rate. Multiple skin grafts were attempted to reduce the fluid loss, but to no avail, as they would not take. With his immune system almost completely destroyed, Ouchi's raw and decaying flesh was vulnerable to attack from bacteria and fungi, while even the few immune cells he had left, cell-dissolving hunter-killers known as phagocytes, turned on him and ate away at his remaining tissue.

On the 59th day of this agony, Ouchi's heart stopped three times in 49 minutes, but despite his expressed wish that he not be kept alive to suffer, he was repeatedly resuscitated. The tragic degeneration of his tortured and increasingly nightmarish body was extensively documented in photographs, which are too distressing to reproduce in this book. The pain that he must have endured is impossible to conceive. He was kept alive, against all odds, and apparently all decency and common sense, until 21 December, the day after his colleague had left hospital. After 83 days, his heart failed and the doctors did not try to resuscitate him. Given that Ouchi himself had reportedly pleaded as early as the first week of his ordeal, "I can't take it any more . . . I am not a guinea pig", how can this extraordinary saga be explained?

Many accounts impute dark motives to the physicians who fought so hard to keep Ouchi from dying. It is widely suggested that the interests of the dying man were callously disregarded by cold-hearted monsters interested only in what appeared to them as a scientific curiosity; that his ordeal was an experiment, prolonged to gather as much data as possible about radiation sickness and the effects of massive exposure. In practice, the story is much less black and white than this, and it seems likely that the primary motivation for Ouchi's treatment came from his family, and their refusal to stop clinging to desperate hopes of a miracle. They spent much of the time camped in an anteroom near him, and they encouraged the doctors to try everything they could. It was only when his family relented and told doctors they wanted him to have a peaceful death that Ouchi was put on a Do Not Resuscitate notice and allowed to pass away.

The chief physician treating Ouchi, Kazuhiko Maekawa, a professor at the University of Tokyo Hospital, spoke afterwards of the complete novelty of the challenge facing his team: "It was the first time I experienced such a case in my 30-plus years as a doctor . . . Almost every day, we came across situations that were not covered in medical textbooks, and there were tough moments in continuing treatment without any sign of a way out." In particular, Maekawa noted that his team were guided by "the attitude of the family . . . Even when the patient fell into critical condition, they never gave up hope and I think that showed how much the family wanted Mr. Ouchi to keep living."

Poor Shinohara followed the same trajectory of hideous decline as Ouchi, over a period of 210 days. He lost 70 per cent of his skin and all his organs failed. Eventually he succumbed to bacterial pneumonia and died on 27 April 2000.

Although Japan's nuclear industry body initially tried to pin the accident on the unfortunate operators, stating, "There is a high likelihood of [the cause being] human error by a worker", it quickly became apparent that the real fault lay in a shocking culture of regulatory complacency and corporate incompetence. The astonishing revelations that followed included the finding that the workers had received almost no training, and that at least one of them reportedly did not even know what the term "criticality" meant. Because the supposed regulators, the Science and Technology Agency (STA), had assumed that a criticality incident was impossible, they had not bothered to make any plans for it. Apparently JCO felt the same, an attitude that was witheringly described by Sixto T. Almodovar, a senior consultant at the Hanford Nuclear Site in Washington state, as "'Titanic thinking': This ship is unsinkable, therefore why obstruct the view of the first-class passengers with unneeded life boats?"

STA inspections were said to have been little more than bureaucratic formalities; inspectors had visited the plant only twice each year – and only once a year in the two years preceding the accident – and never when it was operating. It emerged that workers at the plant had no protective clothing and did not even wear film badges (for monitoring personal exposure to radiation). One expert who visited the plant described it as operating "like a back-street workshop". Almodovar pointed to reports in the Japanese media suggesting that company officials had admitted to encouraging workers to shortcut safety in order to boost productivity, likening it to a practice he had observed at the Y-12 nuclear plant at Oak Ridge in Tennessee: "The Oak Ridge Y-12 workers call this a 'Bubba said'".

JCO headquarters were raided by the police on 6 October 1999, amidst rumours of the existence of a secret manual detailing ways to cut corners and sidestep safety procedures. It was suggested by the *Wall Street Journal* that JCO had been increasingly forced into desperate cost-cutting by the deteriorating economics of the nuclear industry in Japan, where the 1990s had seen escalating public clamour for lower energy prices. In 1996 the Ministry of International Trade and Industry had forced Japanese power companies to cut rates by an average of 4.21 per cent, with another cut of 4.67 per cent following in 1998. As energy utilities looked to maintain their bottom line, they turned to cheaper foreign suppliers for nuclear fuel. The impact on JCO's revenue was dramatic, falling by 48 per cent between 1993 and 1998, and even more severe on the company's profits, which, over the same period, plunged from ¥400 million to ¥97 million.

Eventually JCO conceded that it had violated both normal safety standards and legal requirements, and criminal charges were laid. Its insurers said they would pay out only one billion of the total ¥13 billion liability incurred in the accident; JCO had to fund the rest itself. The year after the accident, the Tokaimura plant's operating licence was revoked, and Japan's nuclear agencies announced a raft of reforms, from enhanced inspection regimes to better training, monitoring and safety equipment and systems. Industry body the Federation of Electric Power Companies (FEPC) insisted the Japanese nuclear industry would become a model of good practice: "We are determined to make every effort to restore people's trust in nuclear safety," said FEPC chairman Hiroji Ota. "Otherwise trust in nuclear power generation as a whole could be undermined." Just over a decade later, that trust would receive a far worse blow than he could have ever imagined, as Japan's "unsinkable" nuclear facilities would once again be cruelly exposed for their "Titanic thinking".

Chapter 11

THE FOUR HORSEMEN: FUKUSHIMA, 2011

Sakiko Araki and her husband Nobikuki lived by the sea, near the Fukushima Daiichi Nuclear Power Plant on the east coast of Japan's Honshu island, about 241 kilometres (150 miles) north of Tokyo. Like most Japanese, Sakiko was well versed in the risks of natural disasters and what to do in case of an earthquake. Living on the low-lying coast facing the Pacific, at the battle front of that ocean's notorious Ring of Fire – the encircling zone of tectonic activity, volcanoes and earthquakes – Sakiko knew an earthquake posed twin dangers, with the threat from the quake potentially followed by the risk of a tsunami. A tsunami is an ocean wave created by displacement of the ocean floor in the event of an earthquake. A shockwave propagates outward through the ocean from above the epicentre of the quake, travelling at speeds of hundreds of kilometres an hour, until it reaches shallow water where its speed and wavelength decrease but its amplitude – the wave height – increases dramatically.

For someone living on the coast of Fukushima Prefecture, a tsunami might be far more deadly and destructive than the quake that caused it, and Sakiko was prepared. She had a grab bag ready at all times and a plan. So when her home was rocked by a massive earthquake early in the afternoon of Friday, 11 March 2011, Sakiko knew what to do. As soon as the shaking had stopped, she started packing the car. Her husband, on the other hand, did not seem to share her sense of urgency; he went inside to get changed, so that he could start clearing up the house. Nobikuki didn't hear the emergency warnings from the roadside speakers warning that a tsunami was coming. "My whole mind was preoccupied with fixing the house," he said. "I didn't think it was going to be so serious . . ." It was only his wife screaming at him, "Forget about the house! We have to go!" that roused him. "If not for her, I would have died."

With Sakiko driving, edging the car between large cracks in the shattered road, they made their escape. Looking back, Nobikuki saw a spray of water rising high into the air over the pine trees along the coast, as a monstrous wave hit the shore. A few miles further up the coast, that same wave was flooding over a seawall guarding a massive nuclear power complex, wiping out electrical lines, blasting away cooling pumps, flooding emergency generator basements and trapping two men in the bowels of a turbine hall.

They did not live to see a succession of calamities overtake the plant, destroying one reactor hall after another and unleashing a nuclear nightmare on Japan.

The Great East Japan Earthquake of 11 March was the strongest ever recorded in that country, with a magnitude of 9.0. Centred below the seabed, 130 kilometres (80 miles) offshore from the city of Sendai in Miyagi Prefecture, it was a rare double quake that lasted for 3 minutes. A 650 kilometre- (404 mile-) long area of the sea floor moved 10–20 metres (32–65 feet) horizontally, giving Japan a mighty tectonic wrench that moved it roughly 2.5 metres (8 feet) east, towards America, and causing the already low-lying coastline of the Miyagi-Fukushima area to subside by over half a metre (1 foot 6 inches). The quake was so violent it shifted the Earth's axis by nearly 10 centimetres (4 inches). The shockwaves from the quake travelled through the Earth's surface at the speed of sound and hit Japan at 14.46. Possibly the first part of mainland Japan to be hit was the Onagawa Nuclear Power Plant in Miyagi Prefecture. About 22 seconds later, the quake hit the Fukushima Daiichi plant.

After the Second World War, and with limited energy resources of its own, Japan had embraced nuclear energy. Work started on the first Japanese nuclear power station in 1961, and by 1991 the country had 54 reactors, generating a third of its electricity. Ten of these reactors were sited at two plants in Fukushima Prefecture – Fukushima Number One and Fukushima Number Two, or, in Japanese, Fukushima Daiichi and Fukushima Daini. The plants were owned and run by the Tokyo Electric Power Company, known as TEPCO. The former, Fukushima Daiichi (F1), with six reactors generating a combined 4.7 gigawatts (GWe), was, in 2011, one of the 15 largest nuclear power stations in the world. All six of the reactors are General Electric design boiling water reactors (BWRs), relatives of the SL-1 in Idaho (see page 144), and distant descendants of the water-shielded core experimental rig at Los Alamos (see page 37). A BWR is a like a huge nuclear kettle, where the core is the heating element, boiling water into steam, which drives turbines to generate electricity. The basic design of the reactors dates back to the early 1960s, and those at F1 came online between 1971 and 1979. Reactor Unit 1 had an output of 460 MWe; Units 2–5, 784 MWe; and Unit 6, 1,100 MWe.

Like all Japanese nuclear plants, F1 had been built next to the ocean so that seawater could be used for cooling, instead of having to build expensive cooling towers. Another major difference from American reactors, is that unlike US PWRs, the BWRs at F1 did not have containment buildings. Whereas PWRs, such as Three Mile Island (TMI), had two cooling loops – a primary one that cooled the reactor, and a secondary one that picked up heat from the primary loop to produce the steam for the turbines – the BWRs at F1 had just one. The cooling loop running through a BWR drives the turbines directly. The steam separators that make this system work sit atop the pressure vessel, making the assembly too tall for a containment building, especially when taking into account extra rigs such as the travelling crane for swapping out fuel. Instead the reactor pressure vessel (RPV) sits inside a dry well: a concrete-encased steel cylinder in the shape of an inverted light bulb. Rather than having the inside of the containment building in which to vent overpressure from steam explosions, the dry well is connected to a toroidal wet well by pipes that come out under water; the idea is that steam is cooled and condensed when it passes through the water, and the water from the torus can then be pumped back into the reactor to cool it. This combined containment structure – known as Mark I containment – is significantly smaller than the containment building of a PWR such as the one at TMI.

Mark I containment was the object of considerable controversy in the 1970s and 80s, with demonstrations and simulations showing how it might struggle to deal with extreme circumstances. The containment structures of the five out of six reactors at F1 that used Mark I containment, had been strengthened and modified extensively. In addition, they were equipped with many safety and backup systems.

The first line of defence was the automatic scram, activated by, for instance, tremors from a quake. In case of a reactor shutdown the turbine would automatically disconnect from the reactor coolant circuit. As at TMI, this immediately removed the major outlet for the reactor's thermal energy, so there would be an urgent need for cooling systems to deal with the heat still being generated by the reactor. Again, as at TMI, this would primarily be the decay heat of the core. Unit 1 was equipped with an isolation condenser, a gravity-driven passive cooling system that circulates hot water/steam from the reactor through a condenser/cooler, in which pipes pass through a tank of cold water. This water absorbs the heat from the reactor water and boils away as steam. There was enough water in the tank for, in theory, three days, but there were connections allowing it to be refilled with fire hoses.

All of the F1 reactors had high pressure coolant injection systems (HPCI). This type of system does not need external power supply, as it is supposed to run off a steam turbine powered by steam from the reactor, which pumps water from the wet well torus into the reactor at up to 16,000 litres (4,250 gallons) per minute, depending on the reactor. However, in order for the HPCI to work, a number of valves must open, and these depend on electricity for their operation. Similarly, electric-motor operated valves are integral to the operation of the reactor core isolation cooling system (RCICS), a safety feature of units 2, 3, 4 and 5, which is similar to an HPCI and takes water from a special reserve tank. Units 2–6 were even equipped with an emergency backup to the emergency backups: a system that pumped seawater to remove residual heat from the reactors, but again, the pump needed electricity to function.

All of the complex valves and almost all the safety systems depended on electricity, to be supplied from a number of potential sources: neighbouring reactors, assuming they were still functioning; the external power grid; one of the twin backup diesel generators with which each reactor was equipped; or, as the absolute last line of defence, a bank of fully charged lead-acid storage batteries, with the capacity to supply power for up to eight hours of emergency operation, although only to the control room systems and valves, as pumps will not run off the batteries. However, several of the systems were designed so that they would run off steam from the reactor, once the valves had been opened.

The guiding principle behind all these backups and safety systems was the "design basis accident": the theoretical worst-case scenario against which the designers needed to guard, and which the plant was engineered to withstand. For instance, with earthquakes a clear and present danger to Japanese plants, F1 had been diligently quake-proofed from the ground up, with the site bulldozed clear of soil to expose the underlying bedrock, so that the reactors could be rooted directly into it. Despite the severity of the 11 March earthquake, the plant would cope well with the initial ground shocks. Unfortunately, the design basis accident did not give sufficient account to the risk of a tsunami, an oversight that would be dissected in forensic detail in the aftermath of the disaster. As a result, the backup generators, battery rooms and electrical switching gear were mostly located in the basement of the plant. At the time of the accident, there was just one backup diesel generator not in a basement – at Unit 6, the newest and most advanced reactor. Protecting all this gear against the threat of a tsunami was a sea wall 10 metres (32 feet) high.

On 11 March 2011, Units 4–6 were offline for refuelling and maintenance; Unit 4 had had its fuel removed, an operation that involved flooding the refuelling floor above the reactor, which is located in a pit in the centre, so that it became a fuel storage pool. This fuel was still generating heat, and so the fuel pool also needed cooling by pumps circulating water through it. Units 5 and 6 had been reloaded with fuel but not yet restarted.

Vivid eyewitness testimony of the moment the quake hit was provided to the Independent Investigation Commission on the Fukushima Nuclear Accident, by an anonymous subcontractor for TEPCO, one of 6,413 workers on site that day:

> . . . the asphalt began to ripple, and I couldn't stay on my feet. In a panic, I looked around and saw a 120 meter exhaust duct shaking violently and looking like it would rupture at any second. Cracks began to appear on the outside of Unit 5's turbine building and on the inside of the entryway to the unit's service building. The air was filled with clouds of dirt.

Like most Japanese (except, apparently, the plant's designers), the workers at the plant were well informed about the dangers attendant on an earthquake for those near the coast. As soon as the quake was over, more than 200 workers ran to the plant entrance, but for security reasons there was a gate with a metal detector, and they could not get out unless the security guard unlocked it. "Let us out of here," they yelled. "A tsunami may be coming!" To their anger, the guard demanded that they, "Wait for instructions from the radiation safety group". His jobsworth attitude was especially galling because many of the workers were aware that in a previous parallel scenario, when an earthquake had struck the Kashiwazaki-Kariwa Nuclear Power Plant in 2007, some workers had faced criminal charges after jumping over the gate to flee.

The subcontractor was dismayed by the damage that the quake had inflicted:

> . . . numerous windows on the second floor [of the operational HQ] had shattered, and the blinds were flapping about in the wind. Three or four cooling towers on the roof had either fallen or were tilted over. Considering that the walls of the newly constructed Units 5 and 6 had been damaged, I figured that Units 1 through 4, which were older, must have been in even worse shape.

In practice, however, although the tremor did rip out some pipe runs and, crucially, knocked down external power lines, the plant came through the quake fairly well. Emergency generators kicked in quickly, and the automatic scram successfully shut down the three operating reactors. In fact, the isolation condenser on Unit 1 worked so well that operators feared it would produce a destructive vacuum by condensing steam too rapidly. Fatefully they started to interfere with its automatic operation, closing two crucial valves.

Meanwhile, out at sea over the epicentre of the quake, the sudden displacement of the ocean floor had also displaced a vast column of water, generating a series of three major tsunamis, which radiated outwards at up to 800 kilometres/hour (500 miles per hour). The first was relatively minor but the second and third, one following hard on the heels of the other, were colossal. They hit the coast of Japan about 50 minutes after the quake. Around 560 square kilometres (216 square miles) of the coastal region were inundated, with a death toll of 19,000 and over a million buildings destroyed or badly damaged.

The subcontractor had made his way to the plant Crisis Centre on the second floor of the earthquake-resistant operational headquarters; it was "jam-packed". With emergency shutdown procedures apparently well in hand, staff at F1 were more worried about Onagawa, up the coast, as they watched the television news showing footage of the tsunami sweeping ashore in Miyagi Prefecture. "But then a section chief came rushing up to Fukushima's plant manager, Masao Yoshida," the subcontractor recalled, "and reported: 'A tank [has] been washed away and had sunk into the ocean.' We all went pale with shock".

The three tsunamis had hit the F1 seawall; the first was relatively low, but the second and third, arriving at 15:35, 49 mins after the quake, with estimated heights of up to 15 metres (49 feet), overwhelmed the wall and flooded the plant. Two engineers trapped in the basement of the turbine building at Unit 4 were drowned; their bodies were recovered three weeks later. According to a report in the *Mainichi Daily News*, they had been ordered to go down into the basement to check for leaks, despite warnings that large tsunamis were inbound.

The tsunami collapsed the water intake structures for all six reactors, wiped out water pumps around the plant, shorted electrical connections, knocked out all the underground diesel generators and cut off the emergency AC power supply. Only the single above-ground generator survived, providing power for Units 5 and 6. Units 3 and 4 still had DC power from the batteries, allowing operators to continue to use the control room, but the battery room serving Units 1 and 2 was flooded and the shared control room went dark. The RCICS for Unit 2 was still running, as it was at Unit 3, but Unit 1 had no emergency cooling of any sort. The passive isolation condenser had been shut off when the valves were closed, and with no power available there was no way to reopen them and get it restarted.

It was about an hour after the scram had shut down fission on the three active reactors. Due to decay heat, they were still producing about 1.5 per cent of their maximum thermal output: about 22 MW in Unit 1 and 33 MW in Units 2 and 3. How long would the emergency systems on Units 2 and 3 hold out? Without cooling, how long would it take for Unit 1 to boil dry and expose the core?

The anonymous subcontractor related a scene of chaos in the operations room: "People continued coming in and out of the Crisis Center, delivering one report after another to Yoshida. Each time, the plant manager's shouts reverberated through a microphone: 'That's not the question I asked!' and 'Give me the answer to . . . this and that!'" But providing the answers was either difficult or impossible. Lighting was out in two of the three control rooms, and it was getting dark outside. Staff were ordered to take desperate measures. "Shortly after 16:00," reported the subcontractor, "we received instructions to 'gather whatever you can,' including hoses, small pumps used for construction work, and emergency light-oil power generators, to help drain water from the electric power room that had flooded. Because we had lost all electric power, it was too dark to get to the electric power room."

This was just the beginning; conditions were getting worse. An International Atomic Energy Agency (IAEA) report on the disaster summed up the situation: "During the initial response, work was conducted in extremely poor conditions, with uncovered manholes and cracks and depressions in the ground. Work at night was conducted in the dark." By rigging up car batteries to supply a few minutes of emergency power to the control room consoles, workers were able to get a snapshot of conditions in the Unit 1 reactor. It was not encouraging. "The water level has begun to fall, and we can no longer see the meters," reported one team. "We can't assess the water level," warned another. "If the water level continues to fall at this rate, the fuel will be exposed by 10 p.m." This was alarming. "All personnel not engaged in work activities," Yoshida ordered, "please evacuate." But the staff in the Crisis Centre were committed. "Nobody got up to go home," recalled the subcontractor. "There was a sense that something had to be done, and it was not an atmosphere in which people felt like rushing off."

Later analysis revealed that by around 17:45 the Unit 1 reactor had already boiled dry; it was too late to avert disaster. But the Crisis Centre had no way of knowing this. In the absence of functioning instruments, they could glean only ominous clues about the worsening plight of the reactor. Just after 19:00, an operator approached the Unit 1 building with only a flashlight to pierce the darkness. Entering the airlock, he approached the inner door and shone his light through the window. "Billowing white steam filling the space on the other side of the glass window", he reported, thinking to himself, "That's raw steam! . . . I had an instinctive sense that going any farther could be very dangerous, so I immediately turned back." When he reported back to the Crisis Centre, there was an explosion of chatter: "What are we going to do? . . . It's not going to explode, right?"

They were right to be worried. By around 19:15 the zircaloy cladding of the fuel elements, heated until it glowed red, began to break up and fall apart, dropping fuel pellets into the bottom of the reactor vessel. Even worse, it started to oxidize, stripping oxygen from the steam molecules to generate hydrogen gas. This reaction is exothermic, meaning it generates heat, causing a positive feedback loop. The possibility of this runaway self-heating meant that overheated fuel assemblies might catch fire even in a fuel pond, which was to become a source of anxiety later on. In the stricken core of the Unit 1 reactor, the exothermic oxidization exacerbated the overheating still further, so that at the bottom of the RPV a combination of fuel pellets, oxidized zirconium and melted control blades began to congeal into corium lava.

Sometime after 20:00, a TEPCO subcontractor showed up at the control room for Units 1 and 2, with more car batteries and small power generators. He reported that

> . . . it was pitch black; a worker had a flashlight in his hand and was trying to read the meter. When I arrived, everybody was ecstatic: "At last, we have light!" We connected the battery directly to a terminal on the back of the meter's control board and began to read the meter.

But the readings seemed to make no sense. A government investigation committee later reported that, at 21:19, the reading for the water level in Unit 1 was "top-of-active-fuel + 200 mm" – in other words, showing that the RPV was filled to a high level. The committee concluded that "it is likely that the meter was not functioning properly". The subcontractor who had witnessed the quake and reported from the Crisis Centre put it more bluntly: "It had become impossible to trust the numbers."

Meanwhile outside the Unit 1 RPV, the immense pressure from the build-up of steam and hydrogen overwhelmed the design tolerance of the dry and wet well structures, splitting open the 2.5 centimetre/1 inch-thick steel containment vessel and filling the reactor building with a cloud of steam, hydrogen and fission products, which mixed with the oxygen in the air inside the building to create a highly explosive mixture. Even without proper monitoring instruments, those in the Crisis Centre could tell that disaster was approaching. Steam was venting from the reactor to the turbine building; elevated radiation levels had been detected as far away as the outer walls of the central control room and areas outside of normal radiological control, indicating that "the volume of radiation was extremely high". "Well, that's the end of this nuclear plant," the subcontractor remembered thinking. "And this is the end of TEPCO."

TEPCO headquarters sent through instructions: vent the reactor building and inject water into the pressure vessel. The only source of water that seemed to be available was from a fire truck that had been on site that day, performing training drills, and which had survived the tsunami unscathed. But no one could work out how to get the water from the truck into the reactor. "We have no hoses!" complained the staff. "We have no plugs! We have no fuel!" Yoshida got on the direct line to TEPCO and told them to send whatever liquid they could find, "It doesn't matter what!" Those in the Crisis Centre knew what venting the reactor building would mean: spreading radioactive contamination into the environment, and on to surrounding communities. A dejected

TEPCO employee in the room muttered, "Well, this is the end of our company."

Nonetheless, venting was obviously essential to relieve the pressure in the building. However, the valve to the vent stack was closed and there was no power to open it. It would have to be operated manually. On the first floor of the operations building, a group of workers, primarily TEPCO employees and people from affiliated companies, formed into response brigades. They organized themselves into five teams of about 20 people each, and members of the radiation-control group were outfitting them with protective suits. The subcontractor from the Crisis Centre was particularly impressed by a young woman who was helping to tape up the seams of the suits. "'She is really dedicated,' I thought: She had asked to stay behind and help." Still more striking was

> . . . the expressions on the faces of the employees assembled into those response teams. Their faces, in the face of lethal danger, were white as sheets. They couldn't find words to express how terrified they were. Every single one of them was trembling; they were truly scared. Nobody knew what could happen. Needless to say, there was a chance they could die.

Final permission to vent did not arrive until 09:03 the next day. Manually operating the valve required workers to sprint through a maze of corridors and doorways, work on the valve, and then run back, all while being exposed to a high radiation field. Meanwhile remaining staff were being evacuated from the site. The subcontractor from the Crisis Centre received his instructions: "Cover your mouth and get on the bus as quickly as possible." Through the windows of the bus, he watched members of the Self-Defense Force (the Japanese military) arrive, even as radiation levels outside continued rising. "They were not wearing masks," he observed. "As the bus drove away, I could not help but think, 'I wonder if those guys are going to be all right.'"

It was now Saturday, 12 March. Over the course of the day, workers managed to pull off a number of difficult tasks, opening the stack valves to vent Unit 1, connecting power to the pumps at Unit 2 and connecting fire hoses to the condenser tanks for Units 1 and 2. Perhaps they allowed themselves a moment's satisfaction; it would be short-lived. At 15:36, the hydrogen in the Unit 1 reactor building exploded, blasting radioactive debris across the plant, injuring five men and undoing much of the painstaking work achieved over the preceding few hours of hard labour. With the plant now contaminated with radioactive debris, all workers at the plant would now require radiation suits and respirators. Having evacuated all but 50 of the staff from the site, Masao Yoshida was now forced to recall evacuated personnel.

They returned to find themselves engaged in a losing battle. At 02:42 on 13 March, the HCPI at Unit 3 finally failed. By 05:30 the fuel in Unit 3 began to collapse and melt, generating hydrogen gas and causing a similar overpressure risk to the one that had just threatened Unit 1. A few hours later, operators managed to open the vent valve and relieve the pressure, but just as with Unit 1, the hydrogen gas was still collecting in the building, and at 11:01 on the 14th, the building exploded, injuring 11 workers, and damaging portable generators and fire engines. Radiation levels at the plant climbed alarmingly. The emergency dose allowed was 100 millisieverts (mSv) (10 rem) per worker. In the yard of the plant, the dose rate was 10 mSv (1 rem) per hour, while at the airlock at the entrance to Unit 3, it was 300 mSv (30 rem) per hour. A worker would be able to spend only 20 minutes there before having to leave the site.

At Unit 2, firefighting equipment had been connected to allow injection of water to take over when the RCICS ran dry, but the Unit 3 explosion destroyed the pumps and hoses. At around 13:00, the RCICS at Unit 2 failed, having worked for 70 hours, and the reactor boiled dry. By 16:30 the fuel assemblies were melting, and yet more hydrogen was being generated. Japan's Nuclear and Industrial Safety Agency (NISA) estimated that 800–1,000 kilograms (1,764–2,205 pounds) of hydrogen had been produced in each of the units. The explosion of Unit 1 had blown holes in the roof of the Unit 2 reactor building, so there was no build-up of hydrogen at the top of the building, although there was later some sort of breach in the containment below the RPV, which caused extensive release of radionuclides. Unfortunately, a cloud of hydrogen from Unit 3 had passed, by way of their shared venting stack, into the upper floors of the Unit 4 building, and at 06:14 on 15 March, it exploded. This caused extreme consternation because the plant managers had no idea where the hydrogen had come from, and assumed that the Unit 4 fuel pond must have been damaged, sustained a leak and run dry, causing the stored fuel rods to become exposed, overheat and generate hydrogen. In fact, the fuel pond, and the stored fuel, were fine.

All the while, workers battled to maintain a supply of cooling water to the reactor cores – or what was left of them. It later transpired that the corium of Unit 1 had melted through the bottom of the RPV and eroded about 65 centimetres (25½ inches) into the concrete of the dry well below (which is 2.6 metres/8 feet 6 inches thick). Some of the corium in Unit 3 was also understood to have melted through the bottom of the RPV and into the dry well. Perhaps the molten cores would never have solidified at all, and kept on going in a China Syndrome scenario (see page 244), but for a brave and controversial decision by the general manager Masao Yoshida. Desperate for coolant, and unable to source enough, Yoshida ordered seawater to be pumped into the reactors. Seawater injection into Unit 1 began at 19:00 on Saturday the 12th, into Unit 3 at 13:12 on the 13th and Unit 2 at 19:54 on the 14th. Famously, on the 12th, Yoshida disregarded orders to stop using the seawater from both TEPCO HQ and the Prime Minister's office; the instruction was withdrawn shortly afterwards.

The cores were not Yoshida's only headache. Without power to run their cooling circuits, the fuel and spent fuel storage ponds on site were also overheating and evaporating away. Of particular concern were the fuel rods in the pool of Unit 3, which contained significant quantities of plutonium. In order to make up the lost water, TEPCO initially had to resort to dropping water by helicopter, and later repurposed a concrete-pumping truck into a water-supply unit.

By the time helicopters were dumping water into the pools, however, most of the drama was over at F1. Almost everything that could go wrong, had gone wrong. With the pumping of seawater stabilizing the situation, and buying enough time to rig up freshwater supplies, cooling of the ruined cores could finally be achieved. Hundreds of cubic metres of water were being pumped into the site every day, rapidly storing up problems for the future (see below). Electrical power to the plant was mostly restored by 22 March. The temperatures of the cores were, finally, coming under control, although TEPCO would hold off declaring a "cold shutdown condition" until mid-December. Events inside the plant, however, were only part of the story.

As disaster mounted upon disaster at the plant, so worried regional and national governments ordered progressively more extensive evacuations. Officials struggled with uncertainty over what was happening in the reactors, let alone what might be the consequences in terms of radioactive releases. The point at which the fuel assembly cladding starts to produce hydrogen is also the point at which some of the most concerning fission products are released from containment, notably radioactive noble gases and gaseous iodine, and aerosols of caesium dust. At Fukushima, one area where the Mark I containment performed quite well is that the suppression pool in the wet well, into which the steam channels vented from the dry well, scrubbed out of the vented cloud a high proportion – 99.9 per cent – of the fission product aerosols. Nonetheless, as soon as TEPCO started to talk about venting the reactor buildings, it was apparent that there would be contamination of the surrounding environs.

Evacuations had begun on the day of the tsunami. At 20:50 on the 11th, the governor of Fukushima Prefecture had ordered the evacuation of the towns of Futuba and Okuma, located within a 2 kilometre (1¼-mile) radius of F1. Less than half an hour later, the Prime Minister extended the evacuation radius to 3 kilometres (2 miles), and ordered a shelter-in-place instruction (where residents are told to stay indoors and close doors and windows) for everyone within 10 kilometres (6 miles). On 12 March at 05:44, the evacuation zone was extended to a 10 kilometre (6 mile) radius; that evening, following the first explosion, the zone was extended to 20 kilometres (12 miles), with evacuation orders for over 78,000 people. Another 62,400 people were affected by a shelter order extending out to a 30 kilometre (18 mile) radius.

There was a fierce debate over what should be set as the tolerable level of contamination, not just for who should evacuate, but also for when they should be allowed to return home. Shunichi Yamashita, a radiobiology expert who had worked in Ukraine in the aftermath of Chernobyl, and who was brought in to advise the authorities at Fukushima, recommended a limit of 100 mSv a year; his experience at Chernobyl had convinced him that the health impact of low dose exposure was greatly over-estimated, and that it was much more important for evacuees to return home as quickly as possible. But in the face of conflicting opinions offered by the International Commission on Radiological Protection, the government hedged their bets with a 20 mSv per year limit. Yamashita became a hate figure: "Afterwards, many people complained that I wanted people to stay in dangerous places."

Government guidance, however, was increasingly falling on deaf ears. Nuclear anxiety was gripping the nation. According to Wade Allison, a nuclear and medical physicist at the University of Oxford, and the author of *Radiation and Reason* (2009), "the public had understood that nothing should go wrong with nuclear plants. Absolute safety was assured. So when it seemed that the impossible had happened, there was panic." Official advice made things worse. On 23 March, overblown concerns about contamination of water led the government to advise parents in Tokyo, over 100 kilometres (60 miles) away, not to let their kids drink tap water. Activity rates of 200 becquerels (Bq) per litre had been measured the day before. "Let's put this in perspective," Allison later cautioned. "The natural radioactivity in every human body is 50 Bq per litre – 200 Bq per litre is really not going to do much harm."

On the INES scale of nuclear disasters, Fukushima is rated at 7, and considered the second worst nuclear accident in history after Chernobyl. But there is some question over whether it merits such a high rating. Japan's NISA initially rated it as a 5 (the same level as TMI, an accident with many points of comparison), but the country's Nuclear Safety Commission (NSC) upped this to "major accident" after evaluation of the radioactive releases. At Fukushima, the main nuclides of concern were iodine-131 and caesium-137, together with some caesium-134. To help make such releases comparable, the radioactivity of the major nuclides involved is compared to that of the primary nuclide of concern, iodine-131, and "translated" into comparable figures, to give a total figure of iodine-131 equivalent radioactivity. In such terms, the NISA estimated in June 2011, that the total release at F1 had been 770 petabecquerels (PBq), about 15 per cent of the Chernobyl release of 5,200 PBq iodine-131 equivalent. In August 2011, however, the NSC revised this down to 570 Pbq. The vast majority of this release had come by the end of March 2011.

Of the three main radionuclides released, iodine-131 has a half-life of just 8 days, while caesium nuclides have a half-life in the human body (the time after ingestion before half of it is expelled through natural processes) of around 70 days. This means that the risk from such nuclides falls rapidly, especially when clean-up operations are in hand. So the long-term impact of the contamination may be relatively minor, and even at the plant, the evidence shows that dose rates fell quickly, from up to 300 mSv per hour, near rubble lying outside Unit 3, to a highest level of just 0.15 mSv/hour near units 3 and 4. The majority of the power plant area is now at less than 0.01 mSv/hour. TEPCO boasts today that you can safely walk around 96 per cent of the 3.5 million-square-metre (37.7 million-square-foot) F1 facility wearing just a jumpsuit and disposable face mask.

Similarly, there is a case to be made that the health impact of these releases and exposures have been minor. An IAEA report from June 2011 estimated that 30 workers at the F1 plant have received doses of between 100 and 250 mSv, amounts that would not be expected to cause any immediate physical harm. For context, 250 mSv was the allowable maximum short-term dose for F1 accident clean-up workers through to December 2011, while 500 mSv is the international allowable short-term dose "for emergency workers taking life-saving actions". Three workers were reported to have suffered suspected radiation burns from wading through heavily contaminated water in a turbine basement, while wearing the wrong boots. After four days of hospital treatment they were released and were not expected to suffer any long-term effects.

By the end of 2011, TEPCO had checked the exposure of 19,594 people who had worked on the site since 11 March, and their figures were revised. Of 167 workers who received doses over 100 mSv, just 23 received 150–200 mSv, three workers received 200–250 mSv, and six received over 250 mSv (with doses ranging from 309 to 678 mSv) apparently due to inhaling iodine-131 fumes. None of these doses are high enough to cause ARS. Exposures for the public were an order of magnitude lower. A typical level of background radiation for someone living in Japan is 2.1 mSv per year. As a result of contamination, someone living in Fukushima Prefecture will probably be exposed to around 10 mSv over an entire lifetime.

In May 2013, the UN Scientific Committee on the Effects of Atomic Radiation (UNSCEAR), following a detailed study by 80 international experts, concluded that "Radiation exposure following the nuclear accident at Fukushima Daiichi did not cause any immediate health effects. It is unlikely to be able to attribute any health effects in the future among the general public and the vast majority of workers." A follow-up report in October 2015 said that there was no new information that "challenged the major assumptions of the 2013 report." In September 2018 it was announced that a man in his fifties, who had worked at the power plant and been exposed to radiation, had died from lung cancer that was diagnosed in 2016. His is the only death directly linked to radiation from the F1 disaster.

The evidence is that evacuation has been far more dangerous to health than radiation. Over 150,000 people eventually fled the region in response to the F1 disaster, among them 81,000 evacuees. Among the evacuees, by December 2018, 2,259 "disaster-related" deaths had been reported.

These deaths were attributed to causes including "spiritual fatigue" and "transfer trauma": psychological health burdens resulting from evacuation. Since "radiation levels in most of the evacuated areas were not greater than the natural radiation levels in high background areas elsewhere in the world where no adverse health effect is evident", this led the World Nuclear Association to conclude that "maintaining the evacuation beyond a precautionary few days was evidently the main disaster in relation to human fatalities". Wade Allison, referring to scenes from the clean-up operation, observed that, "digging up children's playgrounds in protective gear isn't only pointless, it's harmful. It scares people . . . The social effects of scaring people about radiation are fatal."

This is not to say that concerns about radioactive contaminants at the F1 site are not real and ongoing. The primary source of concern today is what to do with all the water that has been, and still is, needed to keep the ruined cores from overheating again. When it passes over the core remnants, this water can become contaminated, and it poses multiple problems. Journalist Roger Cheng, visiting the F1 plant in 2019, described each reactor as "a leaky bucket that constantly needs to be filled with water. At the same time, the water from the leak needs to be collected and stored. And there's no end in sight to this cycle".

As well as the large amounts of water that are deliberately pumped into the plant, groundwater from the surrounding mountains flows through the contaminated below-ground levels of the plant. To try to isolate the dirty water in the plant from the groundwater outside it, an innovative system of freezing pipes was set up from May 2013 to create a 1.6 kilometre/1 mile-long ice wall in the ground surrounding the reactor units. Freezing brine is circulated through a network of 1,568 pipes, extending 30 metres (98 feet) below the surface. The frozen wall started operation in March 2016.

Inside the ice wall, the water in flooded tunnels and trenches has to be decontaminated of its worst nuclides before it can be pumped out. Accordingly, new technologies have been developed, comprising an Advanced Liquid Processing System, which can process 2,000 cubic metres (70,630 cubic feet) of water a day, scrubbing it clean of 62 of the 63 radioactive elements it contains. The only remaining one is tritium, an isotope of hydrogen, which forms part of water molecules and so cannot be scrubbed out. This water is being stored in vast quantities; as of March 2020, more than 1 million tonnes was in storage in 960 huge storage tanks. TEPCO warn that they will run out of tank space in the summer of 2022.

Removal of fuel from the site has been another massive undertaking. Taking away the fuel from the various ponds has been ongoing, but is not expected to be complete until 2031. Removal of the fuel debris from the damaged reactors is much more difficult, and a plan to begin the process has been delayed multiple times. In December 2019, a revised TEPCO draft plan outlined a policy of completing the process "in 30 years to 40 years", and in January 2020, the process was supposed to begin with the drilling of holes to allow access for robotic probes to assess the nature and condition of the debris. When drilling the first hole generated unacceptable levels of radioactive dust, however, TEPCO called a halt to proceedings. This means that even the first steps in the process have not been taken, pushing back yet further any putative end date, much to the frustration of Nuclear Regulation Authority Chairman Toyoshi Fuketa. He publicly chided the TEPCO President Tomoaki Kobayakawa. "Decommissioning won't be complete just by clearing up Fukushima No. 1 and leaving [radioactive waste] inside . . . it's paramount that [work] begins now," Fuketa said. "That falls under the responsibility of management."

Was the management also responsible for the disaster in the first place? Not the tsunami, of course, but the failure of the nuclear power plant to cope with it. This is the key question raised by the F1 disaster: should the accident have been prevented? Perhaps the single most shocking detail to consider when answering this question is that in 1967, when they were constructing the plant, TEPCO actually knocked down most of a huge natural seawall that would have protected F1 from disaster. The natural seawall stood 35 metres (115 feet) high, but TEPCO reduced this to just 10 metres (33 feet). Why would they take such an action? The company, and the regulators that approved the move, considered that, rather than tsunamis, it was typhoon storm surges that were the greatest threat to the plant. "Most large waves in this coastal area are the product of strong winds and low-pressure weather patterns, such as Typhoon

No. 28 in February of 1960, which produced peak waves measured at 7.94 metres," stated a TEPCO submission. Cutting back the seawall saved TEPCO a lot of money, making it much easier to deliver materials by sea and to pump the seawater needed for cooling.

As far as tsunamis were concerned, the original design basis tsunami height at F1 was just 3.1 metres (10 feet), based on the tsunami that had hit the Fukushima coastline after the 1960 Chile earthquake, just a few years before planning for the F1 plant began. In 2001, TEPCO dismissed the possibility of any greater tsunami risk in a one-page memo filed to the Japanese regulator, a decision described as "clearly absurd" by Martyn Thomas at the Royal Academy of Engineering in London. This analysis came despite the fact that there were known to have been eight tsunamis in the region with maximum amplitudes at origin above 10 metres (33 feet) (some much more) – on average, one every 12 years. A 1993 quake had produced a tsunami with maximum height at origin of 31 metres (102 feet), while a massive 8.3 magnitude earthquake in June 1896 produced a tsunami with run-up height of 38 metres (125 feet), killing more than 27,000 people. In 1993, a scientific study reported that a major tsunami of over 15 metres (50 feet) in height was a genuine risk at F1, while a 2006 report warned TEPCO that a tsunami could knock out power and heat sink options for the power plant. Yet the regulator, NISA, told TEPCO that they did not need to do anything immediately, even though IAEA guidelines mandated taking high tsunami heights into account. In 2011, discussions between regulators and TEPCO were ongoing, and it is possible that they might have acquired greater urgency in the light of a forthcoming report from the Japanese Government's Earthquake Research Committee. This study, on earthquakes and tsunamis off the Pacific coastline of north-eastern Japan in February 2011, was due for release in April, and detailed the catastrophic effects on the region of a medieval tsunami.

In other words, TEPCO and the regulators should have known better and done more to prepare. The government's Nuclear Accident Independent Investigation Commission (NAIIC), set up to examine the disaster, was damning. In July 2012, their report harshly criticized the government, the plant operator and the country's national culture. The accident was a "manmade disaster", they said, the result of "collusion between the government, the regulators and TEPCO". It said the "root causes were the organisational and regulatory systems that supported faulty rationales for decisions and actions." The NAIIC chairman wrote:

> What must be admitted – very painfully – is that this was a disaster "Made in Japan". Its fundamental causes are to be found in the ingrained conventions of Japanese culture: our reflexive obedience; our reluctance to question authority; our devotion to "sticking with the programme"; our groupism; and our insularity.

The cost of the disaster is turning out to be enormous. In 2014 the Japanese Government estimated it would cost $75.7 billion to fully decommission and tear down the facility. By late 2016, the estimate for the total clean-up cost, including compensation for evacuees and decontamination of the exclusion zone, had quadrupled to $180 billion. The Japanese Government estimates it will take 40 years to decommission the plant. Lake Barrett, a senior adviser to TEPCO, who previously served as acting director of the Office of Civilian Radioactive Waste Management at the US Department of Energy, has described the effort as being "of the magnitude of putting a man on the moon . . . Unless there's an acceleration, I would not be surprised if it takes 60 years or so."

Yet not everyone is convinced that Fukushima should be a name associated so profoundly with calamity. Wade Allison says that society needs a dose of common sense about radiation, and that the human body has a much greater tolerance for it than is recognized. He argues that the radiation from Fukushima "affected no one's health at all" and that what happened afterwards "was a panic, which was not a physical phenomenon, but a socio-psychological one . . . [which] spread around the world and has people behaving irrationally." Allison insists that, "if people were taught about radiation and there was another Fukushima event tomorrow, it ought not to be seen as an event of global significance."

Whatever its global significance, the tsunami and the nuclear incident it caused had a devastating impact on locals.

Sakiko and Nobukiki Araki, the couple who narrowly escaped the tsunami, lost their home to the water – "[There was] nothing left, not even a piece of wood," Nobukiki told the *Washington Post* in late March, 2011 – and their community to the meltdown: 1,200 former residents of Futabamashi, a neighbourhood near the F1 power plant, spent the weeks following the disaster being shunted from one shelter to another. After a month of sleeping on floors, the Arakis were last heard of planning to move in with their son, who lived near Tokyo and worked, ironically, for a company dealing with nuclear waste. Nobukiki was gloomy about the prospects of returning to his home, and to the surrounding smallholding that his family had tended for generations: "I feel I may never set my foot back on the soil".

APPENDIX

Splitting the Atom

THE CLOSING years of the nineteenth century witnessed an extraordinary flowering of atomic science. The atom was a concept with its roots in antiquity, conceived by Greek philosophers who argued that matter can be divided only up to a point. At the very smallest scale must be units that are indivisible, or atomic. In the early and mid-nineteenth century this ancient creed had been revived as a scientific hypothesis to explain the fundamental nature of matter. Different materials and substances are composed of mixtures and compounds of basic elements; types of matter where each atom of an element is identical to all the other atoms of that element. Atoms of an element have unique qualities that give the element its physical and chemical properties. Precisely how the nature of the atom related to the nature of the element was not yet clear, but in a series of astonishing discoveries at the end of the century, all was revealed.

In 1895, German physicist Wilhelm Conrad Röntgen (commonly spelled Roentgen in English) was experimenting with cathode ray tubes; sealed glass tubes from which air has been expelled to leave an almost complete vacuum, and which have at one end the negative electrode of an electrical circuit, aka the cathode. Previous research had shown that some sort of ray emanates from the cathode, a ray that is able to cast shadows if blocked by items, but which can also drive a paddle wheel to revolve, suggesting that it must be a stream of particles. Roentgen now discovered that a cathode ray tube can generate another mysterious ray, capable of penetrating many types of matter. Since it also exposes photographic plates, this ray could be used to photograph invisible and hidden things, such as bones in a hand. Since the nature of the ray was unknown, Roentgen labelled it with an "X", and duly won the first ever Nobel Prize for physics in 1901, for his discovery of X-rays.

Just a few weeks after Roentgen's discovery, news of it inspired the French physicist Henri Becquerel to investigate whether he could unearth some unknown rays of his own. In January 1896, he began a series of experiments involving photographic plates and crystals of a uranium salt known to be phosphorescent (to radiate light after being stimulated by an external energy source such as a bright light). His original plan involved exposing the uranium salt to sunlight, but one day when it was cloudy he put away the salt and plate combination in a dark box, for storage. To Becquerel's surprise he found that the plate was exposed, even though the uranium salt had not phosphoresced. "I am now convinced that uranium salts produce invisible radiation, even when they have been kept in the dark," he wrote in his lab journal. Becquerel's invisible radiation seemed much less impressive than Roentgen's X-rays, and its true significance was only revealed by a former student of his, Marie Curie. Through a series of brilliant experiments, she showed, as she recalled in an article she wrote for *Century Magazine* in 1904, that "the emission of rays by the compounds of uranium is a property of the metal itself – that it is an atomic property of the element uranium independent of its chemical or physical state." Curie termed this property "radioactivity".

Curie and her husband Pierre went on to discover previously unknown radioactive elements, including radium, laboriously isolated from huge vats of uranium ore, of which it comprises a minute proportion. The tiny speck of pure radium that they managed to obtain promised to overturn millennia of assumptions about the nature of matter. The radioactivity of radium is so intense that it generates enough heat to melt its own weight in ice in an hour. Radium generates this heat continuously, and, as Curie pointed out, "it can be explained by no known chemical reaction . . ." Her conclusion was that

> . . . radioactivity is a property of the atom of radium; if, then, it is due to a transformation this transformation must take place in the atom itself. Consequently,

> from this point of view, the atom of radium would be in a process of evolution, and we should be forced to abandon the theory of the invariability of atoms, which is at the foundation of modern chemistry . . . proof will exist that the transmutation of the elements is possible.

In 1898, further research by New-Zealand-born physicist Ernest Rutherford showed that uranium emits two forms of radiation, which he termed alpha and beta, and Becquerel was able to show that beta radiation comprises a stream of particles with properties identical to the subatomic particle characterized the previous year by the British physicist J. J. Thomson: the electron. Thomson had showed that cathode rays are, in fact, streams of tiny, negatively charged particles, with a mass just 1/1700th that of the smallest known atom, hydrogen.

Over the next three decades, Rutherford and his team put together the pieces identified in the work of Becquerel, the Curies and Thomson, to build a model of the atom that neatly explained many of the fundamental principles of chemistry and physics. The atom is not, after all, indivisible; it is composed of subatomic particles, arranged in a system that, in Rutherford's simplified but still useful model, resembles the solar system. The vast majority of the mass of the atom is contained in a central part, known as the nucleus, around which electrons orbit, a bit like planets around the Sun. Two different types of nuclear particles can be found in the nucleus: positively charged protons, and neutrons, which carry no charge and hence are neutral. In an atom, the positive charge on each proton that is present in the nucleus is balanced by a negatively charged electron in orbit. Thus, a hydrogen atom, which has a single proton, has just 1 electron, while an atom of oxygen, with 8 protons in its nucleus, has 8 electrons. This simple correspondence explains why different elements have different chemical properties. The chemistry of an atom is determined by its electrons; as they make up the "outside" of the atom, they are the parts that interact with other atoms. Adding a proton to an atomic nucleus changes the number of electrons that orbit the nucleus, and this in turn changes its chemical properties. And so it is the number of protons in a nucleus – known as the atomic number – that determines the identity of the element. Hydrogen is hydrogen because it has a single proton, while oxygen is oxygen because it has 8 protons.

But while each and every atom of oxygen must have 8 protons, the number of neutrons in an oxygen nucleus can vary. Since neutrons are neutral, changing the number of neutrons does not affect the electrons and hence does not change the chemistry of the atom. Atoms of the same element that differ in the number of neutrons they have, are known as isotopes. For instance, there are three isotopes of hydrogen: normal hydrogen (sometimes called protium), with no neutrons; deuterium, with 1 neutron; and tritium, with 2 neutrons. All three isotopes have a single proton.

Changing the number of neutrons does not change the atomic number, and so does not change the element, but it does change the mass number of the atom. Different isotopes are identified by their different mass numbers. For instance, the most commonly occurring isotope of carbon has 6 protons (and an atomic number of 6) and 6 neutrons, to give a mass number of 12; it is known as carbon-12. But an isotope of carbon useful for dating organic material has 8 neutrons, and thus a mass number of 14; it is known as carbon-14.

Naturally occurring radioactive elements are those with very large nuclei, and hence with high atomic and mass numbers. The forces that keep nuclear particles stuck together in the nucleus start to weaken at such high mass numbers, and so the nuclei become unstable, and increasingly likely to break apart, or decay. One way they can decay is to spit out an alpha particle, which consists of 2 protons and 2 neutrons (the same as the nucleus of a helium atom). Compared to other forms of radiation, alpha particles are large and heavy; they cannot travel far before smashing into other atomic nuclei, but their size gives them considerable kinetic energy, so they can be very damaging over short ranges. Another way that radioactive nuclei can decay is to emit a beta particle, an electron that is produced when a neutron transforms into a proton. Emitting an alpha or beta particle sometimes leaves a radioactive nucleus in an excited or high-energy state, and such nuclei sometimes shed some of this energy by emitting high-energy photons or electromagnetic waves, known as gamma radiation. All three types of radiation are capable of knocking electrons off other atoms, which creates charged particles known as ions. Hence they are known as ionizing radiation. Such ions tend to be highly reactive and damaging to organic molecules, and so ionizing radiation is deleterious to health. Radioactive isotopes are also known as radionuclides.

As Marie Curie had observed, radioactive decay can produce a lot of energy from a tiny amount of matter. The source of this energy had been revealed by Albert Einstein's equation, $E=mc^2$, which showed that matter and energy are equivalent. Matter is comprised of vast amounts of energy squeezed into a tiny space; radioactivity liberates some of this energy. If a minute quantity of radium can generate an apparently unquenchable and inexhaustible supply of heat, how much energy might be liberated from a larger amount of radioactive material? And what might be the effect if that energy release happened all at once? One of the first to see that this phenomenon would have military implications was the English science-fiction pioneer H. G. Wells, who wrote in his 1914 novel *The World Set Free: A Story of Mankind*, about atomic bombs. Wells described a material where the energy release of one part stimulates energy releases by other parts, which in turn stimulate others: a chain reaction.

Wells' story inspired Hungarian physicist Leo Szilard to wonder whether it would be possible to set off a chain reaction of radioactive decay in a mass of radioactive material. Rutherford felt this was unlikely; radioactive emissions such as alpha and beta particles carry electrostatic charges, and hence they cannot get close enough to other atomic nuclei to destabilize them. When the neutron was discovered in 1932, however, Szilard recognized that it could hold the key. Because it carries no charge, a neutron could approach an unstable nucleus, be absorbed into it, and trigger radioactive decay that might spit out another neutron. He knew that such an event is accompanied by the release of a huge amount of energy. A cascade of such neutron emissions might release an explosive amount of energy. In October 1933 Szilard wrote, "a chain reaction might be set up if an element could be found that would emit two neutrons when it swallows one neutron." A popular analogy is to imagine a field of matches standing upright. Lighting just one match anywhere in the field will generate enough heat to ignite neighbouring ones, and in an instant the whole field will be ablaze.

As well as spitting out types of radiation, an unstable nucleus can undergo more profound changes, breaking apart to give two smaller nuclei. The existence of this phenomenon was first guessed at by the German scientist Lise Meitner and her nephew Otto Frisch, in 1938. Meitner had been involved with research showing that the products of bombardment of uranium atoms with neutrons are two much lighter elements. She and Frisch realized that the uranium nuclei must be destabilized by absorbing the neutrons, so that they split into two, like an amoeba dividing by fission. In 1939 they published a paper about nuclear fission, noting that it releases a tremendous amount of energy. When Szilard worked out that nuclear fission can result in a neutron-release chain reaction, he realized that the phenomenon could be weaponized. It might be possible to build an atomic bomb.

By this time the storm clouds of war were gathering. Both Meitner and Szilard, along with many other scientists, had been driven out of Europe by the Nazis, on account of their religion or politics. Szilard was now working in the United States, where he enlisted the help of fellow emigré Albert Einstein to alert President Roosevelt to the necessity of beating the Nazis in a race to develop nuclear weapons. The result was the setting up of a vast, top-secret research and development project, instituted under the auspices of the Manhattan Engineering District, and codenamed the same. Commonly known as the Manhattan Project, this immense programme pursued several aims in parallel.

Work by Frisch and others (see Chapter 1) had identified that the isotope or radionuclide of uranium most useful as a fissile material – one that can sustain a nuclear fission chain reaction – is uranium-235 (U-235). In naturally occurring uranium ore, U-235 constitutes only a tiny fraction; the vast majority is U-238, a non-fissile isotope. Although U-235 and U-238 are different isotopes, by virtue of atoms of the latter having three more neutrons than atoms of the former, they have the same number of protons and thus the same atomic number, which is why both isotopes belong to the same element. Atoms of both isotopes have 92 protons and 92 electrons, and hence are chemically identical, making it fiendishly hard to separate them. Increasing the concentration of U-235 in uranium samples, known as enrichment, would thus become one of the primary goals of the Manhattan Project.

Another goal of the Project was to show that a fission chain reaction can be started and maintained, and to understand the parameters of this process. One way to do this would be to gather a pile of fissile material. If enough fissile nuclei were present in one place, spontaneous fission of some of them might generate enough neutrons to trigger fission in their neighbours, kicking off a chain reaction. This process, however, is governed by many complex variables. In order for a nucleus to fission, it must absorb or capture a passing neutron. Most of the neutrons produced by a nucleus when it fissions will not be captured by other nuclei; they will escape and be lost. Increasing the likelihood and intensity of a fission chain reaction thus depends on increasing the proportion of neutrons that are captured and trigger fission; this variable is sometimes known as the neutron economy. A non-fissile substance that captures neutrons and thus prevents them from triggering fission is known as a neutron poison; such substances can be used to control fission chain reactions.

One way to improve the likelihood of neutron capture is to use a neutron moderator, a material that can slow down neutrons. Slower neutrons are much more likely to be captured by a nucleus. A good analogy is with a golf ball passing over a hole on the golf course. If it is travelling too fast, it will shoot right over the hole and not go in; but if it can be slowed down, it will fall into the hole. A neutron moderator slows neutrons so that they "fall in the hole"; i.e. are captured by a nucleus. Hydrogen and carbon atoms are good moderators, and so materials rich in these, like water and graphite, are often used as moderators. The human body is also rich in these atoms, and as some unfortunate scientists have discovered to their cost, the simple proximity of a human body can dramatically improve the neutron economy of a fissile system to the point where it becomes dangerous (see page 5, for instance).

Another way to improve the likelihood of neutron capture is to have more fissile nuclei present. The greater the mass of fissile material, the lower the proportion of neutrons that will leak out of the sides of the mass and be lost. The mass of fissile material at which enough neutrons are retained to trigger a self-sustaining fission chain reaction is known as the critical mass, although this term can be misleading, since mass is not the only quantity that matters. More useful terms might include critical assembly or critical system, since these recognize that parameters such as volume and geometry can be important, and that other factors can affect whether the system is critical.

Understanding criticality

A complex but important point is that there are different types of criticality. A subcritical system is one in which each fission, on average, causes less than one new fission event. In such a system, chain reactions might occur but they will quickly fizzle out and will not be self-sustaining.

A critical system is one in which an average of exactly one neutron from each fission causes another fission. This causes a self-sustaining but controlled chain reaction. If this happens when it is not meant to, the event is called a critical accident or criticality.

A supercritical system is one in which each fission, on average, causes more than one new fission. This causes a self-sustaining, uncontrolled, runaway chain reaction with an exponentially increasing neutron population, and liberation of vast amounts of energy in fractions of a second. Supercriticality is what is needed for an atomic bomb. If supercriticality happens when it is not supposed to, it is called a critical or supercritical excursion.

To complicate matters even further, in fission not all neutrons are released at the same time; there is staggered release. For example, when an atom of U-235 splits into two smaller nuclei (for instance, a nucleus of krypton-92 and a nucleus of barium-141), these daughter nuclei, or fission products, are likely to be unstable and radioactive, and so will undergo radioactive decay. When the uranium atom splits, it will immediately – or promptly – release two or three neutrons; these are known as prompt neutrons. The fission products are also likely to release neutrons as they decay, but unlike the prompt neutrons, the release of these will be delayed by anywhere between a few milliseconds to a few minutes after the initial fission event. They are known as delayed neutrons. A system that goes critical due to prompt neutrons is said to be prompt critical; such a system is highly reactive and unstable and is at risk of going supercritical very quickly. A delayed critical system is one that only achieves criticality thanks to delayed neutrons; this is the state that nuclear power generating reactors are aiming for, because it is much more stable, and possible to control.

The world's first nuclear reactor

The first attempt to achieve a controlled fission chain reaction came in 1942, when Manhattan Project scientists assembled a pile of uranium blocks, with graphite moderator, in a squash court at the University of Chicago. This achievement is described in more detail on page 30, but here it is worth describing some of its features, because it was effectively the world's first nuclear reactor. Chicago Pile 1 (CP-1) shared some of the basic anatomy of all reactors. It had radioactive fuel, shaped into short cylinders; subsequent reactors generally use assemblies of pellets of radioactive fuel packed into cylinders, contained in cladding of some sort, and then assembled into rods that can be inserted and withdrawn, the whole being known as a fuel assembly. CP-1 used natural, non-enriched uranium; reactors today generally use uranium that is at least partially enriched. It had moderator, in the form of graphite blocks, machined to provide cavities into which the fuel was packed; some subsequent nuclear reactors have used graphite as moderator, others have used water. CP-1 had control rods, made of sheets of neutron-absorbent cadmium nailed to wooden strips. The default, failsafe set-up was for the control rods to be inserted into the pile; withdrawing them allowed the chain reaction to get going. Control rods or blades have been crucial elements of almost all subsequent reactors.

CP-1 also had an emergency back-up control device: a control rod that hung above the reactor, held in place by a manila rope. A man was stationed to cut the rope with an axe if necessary, so that the rod would drop into the reactor and shut down the fission chain reaction. This back-up device was known as the scram, a term whose etymology is hotly debated. It probably derives from the common slang for getting out of trouble in a hurry. It has been suggested that it was actually an acronym for "safety control rod axe man", although this is generally thought to be a "backronym". Because CP-1 was a low power reactor, it did not possess two features that are essential to all modern reactors: cooling and shielding.

The ultimate aim of the Manhattan Project was, of course, to produce an atom bomb. In fact, two designs of bomb were produced (see next chapter); what they had in common was that both used atomic fission to produce their explosive energy yields. One of the designs relied solely upon fissile uranium, but the other design, which would be used for both the test device detonated in the Trinity test (see page 33), and the Fat Man bomb dropped on Nagasaki, used plutonium as the fissile material. Plutonium is essentially an artificial element that does not occur naturally on Earth, except as transient traces in uranium ores; it is created when a U-238 nucleus captures a neutron, to give U-239, which in turn undergoes two beta decay events that transform two of its neutrons into protons, increasing its atomic number by 2. The resulting nucleus has 94 protons, and has thus transmuted into a new element, named plutonium when it was first produced and isolated in 1940. Isotopes of plutonium such as Pu-239 make excellent fissile material, so that an atomic bomb could be made from as little as 6.2 kg /13 ½ pounds (and theoretically even less) of it. Even as these bombs were being perfected, some in the Manhattan Project had their eyes on a bigger prize; scientists already knew that even greater amounts of energy could be produced via the phenomenon of nuclear fusion, where atomic nuclei are fused together to give larger nuclei. This is the same process that powers the Sun and other stars; on Earth it would be used to produce "superbombs", also known as thermonuclear weapons or, because they use isotopes of hydrogen as the material to be fused, hydrogen or H-bombs.

Understanding radiation units

Multiple units are used to express different aspects of radioactivity, and the picture is confused further because the units used have changed over time. A crucial distinction is between activity, exposure, absorbed dose and equivalent dose (damage done). Activity is the rate of emission of radiation, essentially a measure of how many radioactive decay events take place in a given time, and is a function of the quantity and potency of radioactive material. In the older system, known as the customary system, the unit of activity was the curie, and was based on the rate of radioactive emission of a gram of radium. Supposedly Marie Curie objected to her name being used for a tiny unit, and so this relatively large basis was used, but this in turn has limited the utility of the curie, because a curie of radioactivity is a significant amount. The modern scientific system of units is known as the SI (*Système International*); it is based on the metric system. The SI unit for activity is the becquerel (Bq), but since a becquerel is equivalent to one nuclear decay a second, the becquerel suffers from the opposite problem to the curie, as a single Bq is an infinitesimal amount. One curie equals a colossal number of becquerels: 37 billion of them, or 37 gigabecquerels (Gbq). Radioactive emissions in accidents might be in the range of tera- or petabecquerels (trillions (10^{12}) and quadrillions (10^{15}) of becquerels).

Exposure is a measure of the ionization of air by ionizing radiation – in other words, a measure of the potency of radiation once it has been emitted. In customary units it is measured in roentgens, and in SI units in coulombs per kilogram. Neither of these units features much in this book, because exposure in the technical sense is no longer regarded as a particularly useful metric. Since the 1940s, it has been recognized that what is much more important is the absorbed dose of radiation, a quantity that depends on the type of radiation being absorbed and the type of material absorbing it. Exposure is converted to absorbed dose by multiplying it by a conversion factor, known as the F-factor. For soft tissue the F-factor is very close to that of air, so the F-factor is roughly one, making the conversion easy. The customary unit for absorption is the rad, while the SI unit is the gray (Gy). One gray equals 100 rad.

Radiobiologists are most interested in the biological effects of the dose that is absorbed, and to measure this they use units of equivalent or effective dose. The effective dose measures the exposure of an individual's whole body; the equivalent dose measures the exposure of an organ or tissue. Different organs and tissues have different degrees of sensitivity to radiation; for instance, the spleen, which has rapidly dividing cells, is much more sensitive to radiation than the brain. The effective dose is thus supposed to account for and balance out the different sensitivities of different organs and tissues. The customary unit of effective dose is the roentgen equivalent man or rem, which is still used in the USA, but the SI unit more commonly used elsewhere is the sievert (Sv). One sievert equals 100 rem. Sieverts and rem express how much biological damage has been done by radiation. Since a sievert is a relatively high dose, it is very common to use millisieverts (mSv): thousandths of a sievert.

Different doses are deemed acceptable for the public, those working in the nuclear industry, and those dealing with emergency situations, and this assessment has changed dramatically over time. In 1934 the USA adopted 0.1 roentgens per day (~37 per year) as the standard for whole-body exposure; this is roughly equivalent to 33 rem/year, or 0.33 Sv/year. By 1946, the standard was half of this, and by the late 1950s it was down to 5 rem/year (0.05 Sv or 50 mSv/year). Today a similar level still applies to those who work in nuclear settings. In France, for example, the one-year effective dose limit for people who work with ionizing radiation is 50 mSv, and 100 mSv over 5 consecutive years. In the UK it is 20 mSv in a calendar year, and in the USA it is 50 mSv (5 rem). Compare this to the 250 mSv (25 rem) official maximum dose allowed by the Soviet authorities to workers involved in the clean-up at Chernobyl in 1986 (which was routinely exceeded anyway), while 500 mSv (50 rem) is the international allowable short-term dose "for emergency workers taking life-saving actions".

The standard for the general public is lower still. The International Commission on Radiological Protection (ICRP) recommends an annual effective dose limit of 1 mSv for the general public, excluding doses received from natural background radioactivity or medical exposure (usually X-rays). To put this in perspective, the average total effective dose received by someone living in a country such as France or the UK is 3.7 mSv, of which 2.5 mSv comes from exposure to natural background radioactivity and 1.1 mSv from medical exposure. Just 0.06 mSv of exposure is linked to causes such as radioactive contamination from historical nuclear testing, accidents and normal emissions from the nuclear industry. Cosmic rays bombarding the Earth from space are a type of ionizing radiation, although they are blocked by the atmosphere. At altitude, however, their effects are slightly more potent, so that someone flying from London to New York might expect to receive a dose of 0.032 mSv, the equivalent of a dental X-ray, while someone flying from New York to Hong Kong, over the North Pole, might absorb 0.1 mSv, the equivalent of a chest X-ray. A CT scan might give an effective dose of up to 20 mSv.

The health impacts of radiation exposure depend on whether the exposure is acute or chronic (i.e. short or long term). An acute absorbed dose of 0.7 gray (70 rads) or more will usually cause acute radiation syndrome or sickness (ARS). ARS is characterized by nausea, vomiting, diarrhoea, disorientation, skin darkening (sometimes known as a "nuclear tan"), and subsequently by damage to bone marrow, with concomitant falls in the number of white blood cells in the patient's circulation, and risk of haemorrhage; and of damage to the gastrointestinal and cardiovascular systems. Milder symptoms of ARS can result from absorbed doses as low as 0.3 Gy (30 rads). Effective dose units were not designed to relate to ARS, so absorbed dose units are more meaningful, but roughly speaking, the effective dose threshold for ARS is 100 rem (1 Sv or 1,000 mSv), and a dose greater than 500 rem (5 Sv or 5,000 mSv) is almost invariably fatal.

Effective dose measurements correlate better with long-term health impacts, mainly the risk of cancer, although these are controversial. The conventional assumption is that the risk of illness, such as cancer, is proportional to the dose. These risks are expressed in the context of averages over a population. One way to express this is that a dose of 1 rem (10 mSv) carries with it a 0.05 per cent chance of eventually developing cancer. Potentially more useful is to define collective dose as the sum of the individual doses, to give, for instance, a measurement of person-rem; e.g. 10 people each receiving a dose of 10 rem = 100 person-rem, as do 100 people receiving a dose of 1 rem. A conventional assumption in epidemiology is that, for the general population, there will be one excess cancer fatality per 1,000 person-rem (10,000 person-mSv).

In this book, the units used in each chapter have generally been chosen to match the contemporary usage, since these were the units most used in reports and records of the accidents or incidents. In addition, chapters on incidents involving American nuclear technology follow American practice in using customary units. Note that conversions are usually straightforward, because the conversion factors tend to be multiples of ten, but at the same time it is easy to get confused when magnitude qualifiers such as milli- are used; e.g. 1 rem converts to 0.01 Sv but 10 mSv.

The INES scale

The International Nuclear and Radiological Event Scale (INES) is a tool developed by the International Atomic Energy Authority to help communication between experts and the public. It was explicitly conceived to play a role similar to the

Richter scale for earthquakes by using "a numerical rating to explain the significance of nuclear or radiological events". The scale ranges from 1 ("Anomaly") to 7 ("Major Accident"). Levels 1–3 are termed ""incidents" and levels 4–7 "accidents". Events without safety significance are called "deviations" and are classified Below Scale/level 0. As with the Richter scale, INES is intended to be logarithmic, so that the severity of an event is about 10 times greater for each increase in level on the scale. Assigning a rating to an event involves considering three impact areas: the effects on people and the environment of exposure to radiation or any escape, loss or emission of radioactive material; the performance of radiological barriers and control; and the success of measures for defence-in-depth. This means that an event can rate quite high on the scale even when it had no actual consequences but where the measures put in place to prevent it did not function as intended. INES is not designed for military applications and relates only to safety aspects of events. Using the scale is voluntary, and it is intended for organizations and individuals to use to attribute their own assessments/ratings, rather than these being handed down by a central agency.

CONVERSION TABLE

SI Unit	Customary or Common Unit equivalent
Radioactivity (rate of emission of radiation)	
1 becquerel (Bq) = 1 radioactive decay event per second	
37 gigabecquerels (GBq)	1 curie (Ci)
37 megabecquerels (MBq)	1 millicurie (mCi)
Exposure (rarely used)	
0.000258 coulomb/kilogram (C/kg)	1 Roentgen (R)
Absorbed dose (depends on the exposure and the type of material)	
1 gray (Gy)	100 rad
Effective dose (measure of biological impact of absorbed radiation)	
1 sievert (Sv)	100 rem (roentgen equivalent man)
1 millisievert (mSv)	0.1 rem

Prefixes Often Used with SI Units

Multiple	Prefix	Symbol
10^{12}	tera	T
10^{9}	giga	G
10^{6}	mega	M
10^{3}	kilo	k
10^{-2}	centi	c
10^{-3}	milli	m
10^{-6}	micro	μ
10^{-9}	nano	n